Lecture Notes in Mathematics 2049

Editors:
J.-M. Morel, Cachan
B. Teissier, Paris

For further volumes:
http://www.springer.com/series/304

Angelo Favini • Gabriela Marinoschi

Degenerate Nonlinear
Diffusion Equations

 Springer

Angelo Favini
University of Bologna
Department of Mathematics
Bologna
Italy

Gabriela Marinoschi
Romanian Academy
Institute of Mathematical
Statistics and Applied Mathematics
Bucharest
Romania

ISBN 978-3-642-28284-3   ISBN 978-3-642-28285-0 (eBook)
DOI 10.1007/978-3-642-28285-0
Springer Heidelberg New York Dordrecht London

Lecture Notes in Mathematics ISSN print edition: 0075-8434
ISSN electronic edition: 1617-9692

Library of Congress Control Number: 2012936484

Mathematics Subject Classification (2010): 35K35, 47Hxx, 35R35, 34C25, 49J20

Printed on acid-free paper

Springer is part of Springer Science+Business Media (www.springer.com)

Preface

The aim of these notes is to include in a unitary presentation some topics related to the theory of degenerate nonlinear diffusion equations, treated in the mathematical framework of evolution equations with multivalued maximal monotone operators in Hilbert spaces. The problems concern nonlinear parabolic equations involving two cases of degeneracy. More exactly, one case is due to the vanishing of the time derivative coefficient and the other is provided by the vanishing of the diffusion coefficient on subsets of positive measure of the domain.

From the mathematical point of view, the results presented in these notes can be considered as general results in the theory of degenerate nonlinear diffusion equations. However, this work does not seek to present an exhaustive study of degenerate diffusion equations, but rather to emphasize some rigorous and efficient functional methods for approaching these problems.

The main objective is to present various techniques in which a degenerate boundary value problem with initial data can be approached and to introduce relevant solving methods different for each case apart. The work focuses on the theoretical part, but some attention is paid to the link between the abstract formulation and examples concerning applications to boundary value problems which describe real phenomena. Numerical simulations by which the theoretical results are applied to some concrete real-world problems are included with a double scope: for verifying the theory and for illustrating the response given by the theoretical results to the problems arisen in applied sciences.

The material is organized in four chapters, each divided into several sections. The Definitions, results (Theorems, Propositions, Lemmas), and figures are continuously numbered inside a chapter.

The readers are assumed to be familiar with functional analysis, partial differential equations, and some concepts and basic results from the theory of monotone operators. However, the book is self-contained as possible, some specific definitions and results being either introduced at the first place where they are evoked, or indicated by citations. The work addresses to advanced graduate students in mathematics and engineering sciences, researchers in partial differential equations, applied mathematics and control theory. It can serve as a basis for an advanced course and seminars on applied mathematics for students during the Ph.D. level, and in this respect it is aimed to open to the readers the way toward applications.

The writing of these notes has been developed during the visits of the second author to the Department of Mathematics at the University of Bologna, especially in the periods April–May 2010 when she was a visiting professor, thanks for the financial support of Istituto Nazionale di Alta Matematica "F. Severi"—Gruppo Nazionale per l'Analisi Matematica, la Probabilità e le loro Applicazioni, Italy, May and November 2011. The work is mainly based on the common results obtained with the first author, and some parts of it completed in 2011 are new and not published in other works.

We would like to thank all the reviewers for having lectured this work with obvious patience and carefulness and for all comments, observations, and suggestions which helped to the text improvement.

The authors also acknowledge the PRIN project 20089 PWTPS "Analisi Matematica nei Problemi Inversi" financed by Ministero dell'Istruzione, dell'Università e della Ricerca, Italy and the CNCS-UEFISCDI project PN-II-ID-PCE-2011-3-0027 financed by the Romanian National Authority for Scientific Research, which have contributed to the maintenance of the framework of their collaboration and to the achievement of this work.

Bologna   Angelo Favini
                                                               Gabriela Marinoschi

Contents

Introduction

Before starting the main body of these notes we would like to explain how the equations we shall study arise from real-world problems. Some particularities of these problems can lead to degenerate equations. They involve various interesting mathematical problems whose study will be concretized in general results which, at their turn, can provide useful information while applied to the originary physical problems.

Throughout the work we are concerned with the study of nonlinear degenerate diffusion problems with the unknown function y, consisting basically in the diffusion equation

$$\frac{\partial(u(t,x)y)}{\partial t} - \Delta\beta^*(y) + \nabla \cdot (a(t,x)G(y)) = f(t,x) \text{ in } Q := (0,T) \times \Omega, \quad (1)$$

with initial data and boundary conditions given for the function $y(t,x)$, or in some cases for $u(t,x)y(t,x)$, where the time t runs in $(0,T)$ with T finite and $x = (x_1, \ldots, x_N) \in \Omega$. The domain Ω is an open bounded subset of $\mathbb{R}^N$, with a sufficiently smooth boundary $\partial\Omega$. The boundary conditions can be of Dirichlet type

$$y(t,x) = 0 \text{ on } \Sigma := (0,T) \times \partial\Omega,$$

of Robin type

$$(a(t,x)G(y) - \nabla\beta^*(y)) \cdot \nu = \phi(y) + f_\Gamma(t,x) \text{ on } \Sigma,$$

or of Neumann type if $\phi(y) = 0$. Here, ν is the unit outward normal to Γ. Problem (1) with the initial and boundary conditions can model various diffusion processes in sciences.

As an example we refer to the fluid diffusion in nonhomogeneous partially saturated porous media case in which u accounts for the porosity of the medium, a is a vector characterizing the advection of the fluid through the pores, process also assumed to be influenced by the solution by the means of $G(y)$, and β^* is a (multivalued) function which will be defined a little later in relation with the coefficient of diffusion. The function f expresses the influence of a source or sink distributed in the flow domain. In this case the function y is the fluid saturation in the medium and $\theta(t,x) = u(t,x)y(t,x)$ denotes the volumetric fluid content, or the medium moisture. Another physical process modeled by (1) may be the propagation of a pollutant of concentration y in a saturated porous medium (with all pores filled with fluid). In this application u describes a process of absorption–desorption, namely the retention of the fluid by the solid matrix and the release (of a part) of it after some time. Heat transfer processes are also modeled by equations of type (1). An equation of a similar type is encountered in the models of soil bioremediation or can be deduced under some assumptions from a Keller–Segel chemotaxis model (see [69]). Equations of form (1) with particular coefficients (possibly vanishing) can also characterize nonlinear population dynamics (see [36]), cell growth (see [64]), imaging processes (see [8,9]) and more generally self-organizing phenomena (see [15]). In population dynamics or medical applications y represents the population density.

Some particular properties of β^* place (1) in specific classes of singular diffusion, as we shall see.

In order to justify the physical relevance of the possibly degenerate diffusion problem (1) we shall explain its settlement by giving an example. This will have the role of making clearer how (1), with singular coefficients possibly degenerating in some cases, is deduced from another model (see (2) below) of a physical process. Thus, equations of the more general type (2) can be studied by reducing them to (1).

Let us consider the following equation

$$C(h)\frac{\partial(u(t,x)h)}{\partial t} - \nabla \cdot (k(h)\nabla h) + \nabla \cdot (a(t,x)g(h)) = f(t,x) \text{ in } Q, \quad (2)$$

with an initial condition

$$h(0,x) = h_0(x) \quad (3)$$

and with boundary conditions (of Dirichlet, Neumann or Robin type) which we do not specify at this time. This equation governs the evolution of a certain physical quantity $h(t,x)$.

For instance, (2) particularized in 3D for $g = k$ and $a(x) = (0,0,1)$ is Richards' equations describing the water infiltration into a soil with the porosity u and the conductivity k (see [84], Chaps. 1 and 2). The function h is the pressure in the soil and C is the water capacity. In a porous medium the unsaturated part is the region (or the whole domain) with the pores only partially filled with water, while the saturated part refers to a region where all pores are completely filled with water. By convention, the pressure h is negative in the unsaturated part and positive in the saturated part.

In (2) we assume that C, k and g are nonnegative nonlinear functions defined on (a subset of) $\mathbb{R}$, u and f are real functions defined on Q, u is nonnegative, and a is a vector whose components a_i are real functions defined on Q. The functions C and u are allowed to vanish on a subset of $\mathbb{R}$, or Q, respectively, inducing a certain degeneracy to (2).

In the example concerning Richards' equation we mention that the water capacity C vanishes when the pressure h becomes positive, i.e., in the saturated domain, and it is positive in the unsaturated domain where $h < 0$.

We notice that in the particular case with C vanishing, the time-space domain Q in which the process evolves splits at some time into two subdomains

$$Q_{ns} = \{(t,x);\ C(h(t,x)) > 0\}, \quad Q_s = \{(t,x);\ C(h(t,x)) = 0\}, \qquad (4)$$

separated by a surface whose position is modified in time. The flow in these subdomains is described by a parabolic equation (in Q_{ns}) and by an elliptic equation (in Q_s). Therefore, a degeneracy of (2) induced by the vanishing of C may lead to a free boundary problem. It is obvious that if $C(h) > 0$ for all h, then (2) remains parabolic (if $u(t,x) > 0$) and the free boundary problem does not occur. In our example from soil sciences (4) is a situation in which a simultaneous unsaturated and saturated flow can occur. The subset Q_s is called the saturated domain and Q_{ns} is the unsaturated one, physically corresponding to two phases of the infiltration process.

To treat (2) we shall make some transformations to bring it to the form (1), which is more appropriate to the functional treatment we shall apply.

For the moment let us keep $u(t,x)$ positive.

Let us assume $h_m \in \mathbb{R}$, and

$$C : [h_m, \infty) \rightarrow [0, C_m], \quad C_m > 0,$$

$$k : [h_m, \infty) \rightarrow [K_m, K_s], \quad K_s > K_m \geq 0,$$

$$g : [h_m, \infty) \rightarrow [g_m, g_s].$$

We shall further refer only to some basic and interesting cases from the mathematical point of view. We consider that the functions C, k and g are single valued, continuous and bounded. More generally they may have discontinuities of first order, at most, i.e., at the points where they are not continuous they have finite lateral limits. In this case some modifications will occur in the model and in its mathematical treatment, without essentially changing the arguments.

We define $C^* : [h_m, \infty) \rightarrow [y_m, y_s]$, as

$$C^*(h) = y_m + \int_{h_m}^{h} C(\zeta)d\zeta, \ h \geq h_m, \qquad (5)$$

and $K^* : [h_m, \infty) \to [K_m^*, \beta_s^*]$ by

$$K^*(h) := K_m^* + \int_{h_m}^{h} k(\zeta)d\zeta, \ h \geq h_m, \ K_m^* \geq 0 \tag{6}$$

which are continuous nondecreasing functions, where y_s is a real positive number and $y_m = C^*(h_m)$ is a nonnegative number.

We make the function notation $y = C^*(h)$ and assume first that the medium is unsaturated i.e., C is positive. Then C^* turns out to be monotonically increasing and its inverse is $(C^*)^{-1} : [y_m, y_s] \to [h_m, +\infty)$,

$$h = (C^*)^{-1}(y), \ y \in [y_m, y_s]. \tag{7}$$

Replacing h in (2) this becomes

$$\frac{\partial(u(t,x)y)}{\partial t} - \Delta\beta^*(y) + \nabla \cdot (a(t,x)G(y)) + C_1(y)\frac{\partial u}{\partial t}(t,x) = f(t,x) \text{ in } Q, \tag{8}$$

where the composed functions

$$\beta^*(y) := (K^* \circ (C^*)^{-1})(y), \ y \in [y_m, y_s], \tag{9}$$

$$G(y) := (g \circ (C^*)^{-1})(y), \ y \in [y_m, y_s], \tag{10}$$

$$C_1(y) = C((C^*)^{-1}(y)) \cdot (C^*)^{-1}(y) - y, \ y \in [y_m, y_s]$$

occur. Moreover, we still define

$$K(y) := (k \circ (C^*)^{-1})(y), \ y \in [y_m, y_s]. \tag{11}$$

Now, we shall separate several cases of interest related to the possible vanishing of C on a subset included in (h_m, ∞). As specified before, in the theory of water infiltration in soils $h_m < 0 = h_s$, $y_m = 0$ and $g = k$. At the value $h_s = 0$ the saturation occurs. The function y is called the *water saturation* in soil and the corresponding value $y_s = C^*(h_s)$ is called the *saturation value*.

(a) *The fast diffusion case.* Let us consider that C is continuous

$$C(h) = 0, \ h \in [h_s, \infty), \ \ C(h) > 0, \ h \in [h_m, h_s). \tag{12}$$

Then (2) degenerates on $[h_s, \infty)$ and (5) becomes

$$C^*(h) = \begin{cases} y_m + \int_{h_m}^{h} C(\zeta)d\zeta, & h_m \leq h < h_s, \\ y_s, & h \geq h_s, \end{cases}$$

namely $y = C^*(h)$ remains at the saturation value y_s as well as h is positive. In this case one can compute the inverse of C^* on the interval $[h_m, h_s)$ where it is increasing, while for the interval $h > h_s$ the inverse is no longer a function, but a graph, i.e., at the point $y = y_s$ it has as image the whole interval $[h_s, \infty)$. More exactly we have

$$(C^*)^{-1}(y) := \begin{cases} (C^*)^{-1}(y), & y \in [y_m, y_s), \\ [h_s, +\infty), & y = y_s, \end{cases} \tag{13}$$

and so the function $h = (C^*)^{-1}(y)$ is continuous and monotonically increasing on $[y_m, y_s)$ and so-called *multivalued* at $y = y_s$. Then, by a direct replacement in (6) one obtains a function K^* with two branches

$$K^*(h) := \begin{cases} K_m^* + \int_{h_m}^h k(\zeta)d\zeta, & h \in [h_m, h_s), \\ \beta_s^* + K_s h, & h \geq h_s, \end{cases}$$

which plugged in (9) leads to a multivalued function

$$\beta^*(y) := \begin{cases} (K^* \circ (C^*)^{-1})(y), & y \in [y_m, y_s), \\ [\beta_s^*, +\infty), & y = y_s \end{cases} \tag{14}$$

which has the image $[\beta_s^*, +\infty)$ at $y = y_s$. Here, we took $h_s = 0$ and

$$\beta_s^* := \lim_{y \nearrow y_s} (K^* \circ (C^*)^{-1})(y) > 0. \tag{15}$$

We notice that β^* is continuous on $[y_m, y_s)$ and that $K(y_m) = K_m$, $K(y_s) = K_s$. Since the function $(C^*)^{-1}$ is monotonically increasing on $[y_m, y_s)$ we can calculate $\beta^*(y)$ by changing the variable in the integral (6), by denoting $\zeta = (C^*)^{-1}(\xi)$. In this way we get

$$\beta^*(y) = \begin{cases} K_m^* + \int_{y_m}^y \beta(\xi)d\xi, & y \in [y_m, y_s), \\ [\beta_s^*, +\infty), & y = y_s \end{cases} \tag{16}$$

and now in (8) the sign "=" can be replaced by the sign "$\ni$". In the above expression

$$\beta(y) := \frac{k((C^*)^{-1}(y))}{C((C^*)^{-1}(y))}, \text{ for } y \in [y_m, y_s). \tag{17}$$

The function β defines the *diffusion coefficient* (also called diffusivity for certain physical processes) and under the hypotheses made before it is a nonnegative function, satisfying the blow-up property

$$\lim_{y \nearrow y_s} \beta(y) = +\infty. \tag{18}$$

Concerning detailed computations of the above relations and the various hypotheses made for k which may lead to some specific models we refer to [84], Sect. 2. For a further use we define the ratio

$$\rho := \frac{k(h_m)}{C(h_m)} = \frac{k((C^*)^{-1}(y_m))}{C((C^*)^{-1}(y_m))} \tag{19}$$

which is nonnegative. Let us mention that if C and k are defined on $\mathbb{R}$ (as it happens in some models), i.e., if $h_m = -\infty$, then we replace the above ratio by

$$\rho := \lim_{h \to -\infty} \frac{k(h)}{C(h)}$$

which we assume finite.

We mention that the last term on the left-hand side in (8) is a single-valued function because $C_1(y) = -y$ for $y = y_s$. If u does not depend on t it vanishes.

In our example related to Richards' equation, by defining the functions $y = C^*(h)$, representing the water saturation in pores, β (the water diffusivity) and $G = K$ (the water conductivity function) we have passed from Richards' equation written in terms of pressure to its diffusive form (1) written for the water saturation $y \geq y_m \geq 0$. Generally y_m is taken equal with zero.

In the mathematical treatment we shall need to work with these functions defined up to $-\infty$, so that we extend the β at the left of $y = y_m$ by ρ if $\rho > 0$ and by a continuous positive function if $\rho = 0$. The function K and G are extended by $K(y_m)$ and $G(y_m)$ at the left of $y = y_m$.

To resume, this case is characterized by the functions β and β^* defined on the subset $(-\infty, y_s)$, having singular behaviors at $y = y_s$

$$\lim_{y \nearrow y_s} \beta(y) = +\infty, \quad \beta^*(y_s) \in [\beta_s^*, +\infty). \tag{20}$$

A typical example is

$$\beta(y) = \frac{1}{(y_s - y)^{1-p}} \text{ for } 0 < p < 1.$$

By analogy with the classification given by Aronson in [6] we say that this case defines a *fast diffusion* and models a free boundary process.

Therefore, problem (2), which degenerates due to (12) and has been transformed into (1) with a multivalued function β^* is relevant for a free boundary problem. The fact that a degenerate equation (2) corresponding to a free boundary problem is characterized by an equation with a multivalued operator must not be surprising because the extension of a nonlinear function to a multivalued one by "filling in the jumps" with graphs is common in the theory of nonlinear differential equations with discontinuous coefficients as well as in that modelling free boundary processes (see [14] for various examples).

Under some other assumptions (regarding mostly the discontinuity of these functions) one can get some other particular properties of the functions β, β^* K and G (e.g., they can be multivalued at some points within $[y_m, y_s)$, see [86]).

(b) *The superdiffusion case.* In this case let us allow β_s^* defined in (15) going to infinity, i.e., assume

$$\lim_{y \nearrow y_s} \beta(y) = +\infty, \quad \lim_{y \nearrow y_s} \beta^*(y) = +\infty. \tag{21}$$

The situation in which both β and β^* blow up at $y = y_s$ corresponds to a singular expression of β,

$$\beta(y) = \frac{1}{(y_s - y)^{1-p}} \text{ for } p \leq 0$$

and was defined in [6] as *very fast diffusion*, or *superdiffusion*.

An interesting example from biology is the instantaneous disappearance (due to an extremely high diffusion coefficient) of a population of locusts when its density reaches a certain critical value, y_s.

(c) *The slow diffusion case.* We assume that

$$C(h) > 0 \text{ for } h \in [h_m, \infty) \tag{22}$$

and

$$\lim_{h \to \infty} C(h) = 0. \tag{23}$$

This implies by (17) that $\beta(y) < \infty$ for $y \in [y_m, +\infty)$ and

$$\lim_{y \to \infty} \beta(y) = +\infty.$$

This case which may be illustrated by

$$\beta(y) = y^p \text{ for } p > 1$$

is a *slow diffusion* and the equation with such a diffusion coefficient is known as the porous media equation, because it generally models the diffusion of a gas in a porous medium. It also models nonlinear heat diffusion. If $p = 1$ we get the classical heat equation. Concerning more general studies on diffusion equations we refer the reader to [96–98]. If

$$\lim_{h \to \infty} C(h) \geq C_s > 0$$

then

$$\lim_{y \to \infty} \beta(y) \le \beta_s = \frac{K_s}{C_s} < +\infty$$

and β^* has a sublinear increasing, $\beta^*(y) \le \beta_s y$.

In the cases (b) and (c) the function β^* turns out to be single valued. Case (c) characterizes a unsaturated flow, only.

To conclude this presentation we give some examples of functions C, β, K which are basic in soil sciences. They are empirical relations established by observations. Let us take the porosity constant and by dividing by it we can rewrite (8) with $u = 1$.

First we refer to the hydraulic model of van Genuchten which proposes the hydraulic functions (in a particular case)

$$K(\widetilde{y}) := \begin{cases} K_s^{0.5} \widetilde{y}[1 - (1 - \widetilde{y}^{1/m})^m]^2 & \text{if } \widetilde{y} < 1, \\ K_s & \text{if } \widetilde{y} = 1, \end{cases}$$

$$\widetilde{y}(h) := \begin{cases} [1 + |\alpha h|^{1/(1-m)}]^{-m} & \text{if } h < 0, \\ \widetilde{y}_s & \text{if } h \ge 0, \end{cases}$$

where $m \in (0, 1)$, $\widetilde{y}$ is the dimensionless water saturation defined by

$$\widetilde{y} := \frac{y - y_m}{y_s - y_m}$$

and α is a length scaling factor. Obviously, the dimensionless saturation value $\widetilde{y}_s$ is equal to 1 and $K_s = K(\widetilde{y}_s)$. The water capacity is then

$$C(h) := \begin{cases} \frac{m}{1-m} \left\{ 1 + |\alpha h|^{\frac{1}{1-m}} \right\}^{-m-1} |\alpha h|^{\frac{m}{1-m}} \frac{|h|}{h}, & \text{if } h < 0 \\ 0, & \text{if } h \ge 0. \end{cases}$$

The various values of the parameter m correspond to more or less nonlinear behaviours of the soil. For example, we can notice that if m is close to 0, there is a very low rate of variation of $\widetilde{y}(h)$. The point at which C attaints its maximum is very close to the saturation point and the hydraulic conductivity K evolves highly nonlinear. This approaches a slow diffusion case. If m is close to 1, we have that

$$\lim_{h \nearrow 0} C(h) = 0, \quad \lim_{\widetilde{y} \nearrow 1} K'(\widetilde{y}) < +\infty$$

and we can notice a nonlinear variation of $\widetilde{y}(h)$, and a more linear behaviour of the hydraulic conductivity. This leads to a fast diffusion model with a strongly nonlinear transport.

The parametric model of Broadbridge and White (see [33]) introduces as hydraulic functions the diffusion coefficient and the conductivity

$$\beta(\widetilde{y}) = \frac{c(c-1)}{(c-\widetilde{y})^2}, \ K(\widetilde{y}) = \frac{(c-1)\widetilde{y}^2}{c-\widetilde{y}}, \tag{24}$$

with the same significance as before for $\widetilde{y}$. Here, the hydraulic nonlinearity of the medium is characterized by the parameter c belonging to $(1, +\infty)$. If $c \to 1$ the medium is strongly nonlinear and if $c \to \infty$ the medium behaves weakly nonlinear. We can define

$$\beta^*(\widetilde{y}) = \begin{cases} \frac{(c-1)\widetilde{y}}{c-\widetilde{y}} & \text{if } \widetilde{y} < 1 \\ [1, \infty) & \text{if } \widetilde{y} = 1 \end{cases} \tag{25}$$

and notice that this corresponds to a slow diffusion for $c \gg 1$ and to a fast diffusion when c is approaching the value 1.

These are some examples intended to show that the mathematical properties previously considered for the functions β and K are not formal but they can be retrieved in the usual hydraulic functions in practice. These properties, pretty general, will be transformed into mathematical hypotheses for the boundary value problems discussed in the next chapters.

Without giving an exhaustive information we would like to cite some key achievements in the literature regarding existence and uniqueness studies for solutions to the degenerate equation

$$\frac{\partial(g(h))}{\partial t} - \nabla \cdot (b(\nabla h, g(h))) + f(g(h)) = 0.$$

We refer first to the method of entropy solutions introduced by S.N. Krushkov in [75, 76]. Originally this method was devoted to prove L^1-contraction for entropy solutions for scalar conservation laws, i.e., generalized solutions in the sense of distributions satisfying admissibility conditions similar to those of entropy growth in gas dynamics (see also [23]). H.W. Alt and S. Luckhaus established in [4] various existence, uniqueness and regularity results for initial-boundary value problems for quasilinear systems of the above form and studied variational inequalities of elliptic–parabolic equations applying the results to certain Stefan type problems. J. Carillo (see [37, 38]) applied Krushkov's method to nonlinear equations

$$\frac{\partial(g(h))}{\partial t} - \Delta b(h) + \nabla \cdot \phi(h) = 0$$

with g and b continuous nondecreasing functions and ϕ satisfying rather general conditions. F. Otto (see [92]) proved a L^1-contraction principle and uniqueness of solutions for this type of equation by applying Krushkov's technique only to the time variable. We also mention the works [65, 66] referring to equations with degenerate diffusion operators.

A degenerate, doubly nonlinear parabolic system, with coefficients depending on time and space too, was treated in [70] by approximating it by a nondegenerate one and proving the convergence of approximate solutions.

In the paper [24] a model of the saturated–unsaturated flow lying on a special definition of the boundary conditions that changes during the phenomenon evolution, has been developed also for a finite value of the diffusivity at saturation (which was implied by the assumption that $C(h) > 0$, $\forall h \geq 0$). Following the technique presented in [50] the model was reduced to systems in class of Stefan-like problems of high-order, see [49].

Another aspect which appears to be very relevant for nonlinear diffusion problems refers to the solution dependence on the choice of nonlinear functions involved in the equation. Such a problem in relation with the fluid flow in a porous medium obeying Richards' equation was recently treated by Borsi et al. in [25]. This article deals with the finite-time stability (in fact the continuity) of solutions to nonlinear scalar parabolic boundary value problems in 1-D with respect to variations of parameter functions and gives extensive details about applications in groundwater flows.

The analysis of the well-posedness of (1) in the very fast diffusion case with u constant was approached in the papers [20] (with Robin boundary conditions) and [81] (with nonhomogeneous Dirichlet boundary conditions) in the framework of evolution equations with m-accretive operators in Hilbert spaces. The latter gives also an example of a nonautonomous problem.

A rigorous existence and uniqueness theory for the fast diffusion case with a free boundary occurrence was begun in [82] for $N = 1$ and developed in [83] for $N = 3$, for $u = 1$. The fast diffusion case with the function k (and consequently β) discontinuous was treated in [86].

As another application, we mention a particular degenerate model characterizing diffusion in partially fissured media which was treated in [94] by extending the existence and uniqueness results proved in [48].

To resume, we have established that (1) characterizing a diffusive flow may be deduced by (2) which can be a degenerate equation (due to the vanishing of $C(h)$) and in this case β blows up and β^* turns out to be multivalued at y_s. In this work we shall associate this case with two situations in which (1) still degenerates due to the hypotheses further presented.

The parabolic–elliptic degenerate case. Let us assume that $u(t,x)$ may vanish on a subset Q_0 of Q of zero measure. This leads to a degeneration of (1) into an elliptic equation on Q_0 and we shall call this case *parabolic–elliptic degenerate*.

The diffusion degenerate case. Let $u(t,x) > 0$ and assume that $k(h_m) = 0$. Then, by (17)

$$\begin{cases} \beta(y) > 0 \text{ for } y > y_m \\ \beta(y) = 0 \text{ for } y = y_m. \end{cases}$$

This implies that the diffusion coefficient $\beta(y)$ vanishes on the subset of Q where $y(t, x) = y_m$ and ρ given by (19) is zero. For convenience we shall call this case *diffusion degenerate*.

If $k(h_m) > 0$ the diffusion coefficient β is positive, i.e., $\rho > 0$, and then we deal with a nondegenerate diffusion case.

We indicate the monograph [62] and the articles [2,3,18,19,51–57,61,63] as a few references concerning the theory of linear equations degenerating into elliptic or hyperbolic ones, examples and identification problems in relation to them.

In this book we treat nonlinear equations degenerating in elliptic ones or degenerating due to the vanishing of diffusion coefficient and focus especially on the fast diffusion case which has a special physical and mathematical relevance being associated with a free boundary problem whose character is concisely expressed by the graph β^*.

The set of hypotheses which will be assumed in each chapter are pretty general and follow from the properties of the empirical functions used in practice and envisage especially singular cases. The techniques used to solve the problems in each chapter can be extended to other more general cases including: equations with discontinuous coefficients (see e.g. [86]), K non Lipschitz, hysteretic behaviour of the nonlinear functions (see e.g. [85]).

Since the monotonicity of the main nonlinear term in the equations corresponds to a natural dissipativity assumption for these classes of problems and has an obvious physical meaning, the methods we use are related to the theory of nonlinear evolution equations with monotone (accretive) operators in Hilbert spaces. They provide rigorous and efficient techniques for approaching well-posedness and other mathematical aspects raised by these problems. In few lines we present a short history of the m-accretivity technique. The main results of the theory of nonlinear maximal monotone operators in Banach spaces are essentially due to Minty (see [79, 80]) and Browder (see [34,35]). Further important contributions are due to Brezis (see [26–28]), Lions [77], Rockafeller [93], mainly in connection with the theory of subdifferential type operators. The general theory of nonlinear m-accretive operators in Banach spaces has been developed in the works of Kato (see [72]) and Crandall and Pazy (see [41, 42]) in connection with the theory of semigroups of nonlinear contractions and nonlinear Cauchy problems in Banach spaces. The existence theory for the Cauchy problem associated with nonlinear m-accretive operators in Banach spaces begins with the pioneering papers of Komura (see [74]) and Kato [71] in Hilbert spaces and was extended in a more general setting by several authors. For a complete approach of these problems we refer the readers to the monographs [10,14]. At the appropriate places in the text we shall specify the concepts related to the m-accretivity techniques and indicate the main results to which the proofs make appeal.

Chapter 1 is concerned in several sections with the study of the existence and various properties of the solutions to (1) which degenerates in an elliptic equation. We shall treat this case with a nondegenerate diffusion coefficient. Using techniques from the theory of evolution equations with m-accretive operators in Hilbert spaces results of a higher degree of generality are obtained. Briefly, we shall treat the following problems: existence properties of the solution to (1) in the fast diffusion case, with homogeneous Dirichlet boundary conditions, for u function of x only, vanishing on a subset Ω_0 of Ω with meas$(\Omega) > 0$. In the abstract framework of evolution equations one associates to this case the abstract Cauchy problem

$$\frac{d(My)}{dt}(t) + Ay(t) \ni f(t) \text{ a.e. } t \in (0, T), \tag{26}$$

$$(My)(0) = \theta_0$$

where M is not invertible and A is a multivalued nondegenerate operator. The method consists in an approximation of the operators M and A, the proof of the existence of an approximating solution and a convergence result together with a special construction of a weak solution to (26). Sufficient conditions under which uniqueness follows are established.

The Cauchy problem will be studied further under the hypothesis of a very fast diffusion in which the existence proofs necessitate some different arguments than in the fast diffusion case. They will be detailed in a separate section.

Next, the existence of periodic solutions with f periodic as well as other results concerning the longtime behavior of Cauchy problems with periodic data will be investigated.

In Chap. 2 the problems of existence and uniqueness for the Cauchy problem will be approached in the diffusion degenerate case in which (1) with u constant is accompanied by certain Robin boundary conditions. We shall deal with the abstract problem

$$\frac{dy}{dt}(t) + Ay(t) \ni f(t) \text{ a.e. } t \in (0, T), \tag{27}$$

$$y(0) = y_0$$

where A is a multivalued operator and this case will be approached via a time-difference scheme whose stability and convergence will be studied. The existence of periodic solutions as well as the longtime behavior of the solutions to Cauchy problems with periodic data will be further studied in the diffusion degenerate case, too.

In Chap. 3 a nonautonomous parabolic–elliptic degenerate problem with u depending on t and x is studied. This involves the application of some results of Kato and Crandall and Pazy for semigroups generated by time-dependent nonlinear multivalued operators.

Chapter 4 is concerned with an identification problem, approached as a control problem, for the vanishing coefficient u which will be determined from available observations upon the solution. We resort to an auxiliary approximating problem and determine the optimality conditions for an approximating optimal pair. Then we prove a convergence result of the sequence of approximating control pairs (control and state) to a solution to the original problem.

Some sections are concluded by numerical simulations intended to put into evidence the specific features of the solutions in the context of their physical relevance.

The problems we deal in the book involve mainly nonlinear operators, so that for all concepts, definitions and results related to the nonlinear operators in Banach spaces and semigroup theory we refer the readers to the monographs [10–14, 22, 29, 30, 95, 99].

Chapter 1
Existence for Parabolic–Elliptic Degenerate Diffusion Problems

In this chapter we are concerned with the study of some boundary value problems with initial data formulated for parabolic–elliptic degenerate diffusion equations with advection, focusing especially on the fast diffusion case which involves a free boundary problem (case (a) in Introduction). After setting an adequate functional framework for each situation we transpose the boundary value problems into abstract formulations and study their well-posedness with specific methods of the theory of nonlinear evolution equations with m-accretive operators in Hilbert spaces. We investigate the conditions under which particular properties of the solutions, like uniqueness and time periodicity take place. We mention that the case without advection was studied in [58]. Numerical simulations applied to problems arisen in soil sciences complete the study and sustain the theoretical achievements.

Notation. We specify the functional spaces which will be further used.

Let Ω be a open bounded subset of $\mathbb{R}^N$ ($N \in \mathbb{N}^* = \{1, 2, \ldots\}$), with the boundary $\Gamma := \partial\Omega$ sufficiently smooth. The space variable is denoted by $x := (x_1, \ldots, x_N) \in \Omega$ and the time by $t \in (0, T)$, with T finite.

We shall work with the spaces $L^p(\Omega)$ (see [30], pp. 89), Sobolev spaces $W^{m,p}(\Omega)$ (see [30], pp. 263, 271) and the vectorial spaces $L^p(0, T; X)$, $W^{m,p}(0, T; X)$ where X is a Banach space (see [14], pp. 21), $m \geq 1$ and $p \in [1, \infty]$. Briefly, we recall that

$$L^p(\Omega) = \{f : \Omega \to \mathbb{R}; \ f \text{ measurable, } |f(x)|^p \text{ integrable}\}, \ p \in [1, \infty),$$

$$L^\infty(\Omega) = \left\{ \begin{array}{l} f : \Omega \to \mathbb{R}; \ f \text{ measurable and there is a constant } C \\ \text{such that } |f(x)| \leq C \text{ a.e. on } \Omega \end{array} \right\}$$

A. Favini and G. Marinoschi, *Degenerate Nonlinear Diffusion Equations*, Lecture Notes in Mathematics 2049, DOI 10.1007/978-3-642-28285-0_1, © Springer-Verlag Berlin Heidelberg 2012

are Banach spaces with the norms

$$\|f\|_{L^p(\Omega)} = \left(\int_\Omega |f(x)|^p \, dx \right)^{1/p},$$

$$\|f\|_{L^\infty(\Omega)} = \inf\{C; \ |f(x)| \le C \text{ a.e. on } \Omega\},$$

respectively. For $m \ge 1$ and $p \in [1, \infty]$ the Sobolev space $W^{m,p}(\Omega)$ is defined by

$$W^{m,p}(\Omega) = \{f \in L^p(\Omega); \ f \text{ measurable and } D^\alpha f \in L^p(\Omega), \text{ with } |\alpha| \le m\}$$

where α is a multi-index and $|\alpha| = \sum_{i=1}^N \alpha_i$, α_i is a positive integer and $D^\alpha = \frac{\partial^{|\alpha|}\varphi}{\partial x_1^{\alpha_1}....\partial x_N^{\alpha_N}}$.

The norm is defined by

$$\|f\|_{W^{m,p}(\Omega)} = \left(\sum_{1 \le |\alpha| \le m} \|D^\alpha f\|_{L^p(\Omega)}^p \right)^{1/p}, \text{ if } 1 \le p < \infty,$$

$$\|f\|_{W^{m,\infty}(\Omega)} = \max_{1 \le |\alpha| \le m} \|D^\alpha f\|_{L^\infty(\Omega)}^p, \text{ if } p = \infty.$$

We still denote $H^m(\Omega) = W^{m,2}(\Omega)$ which is a Hilbert space with the scalar product

$$(u,v)_{H^m(\Omega)} = \sum_{1 \le |\alpha| \le m} (D^\alpha u, D^\alpha v)_{L^2(\Omega)}.$$

Let X be a Banach space. We denote

$$L^p(0,T;X) = \left\{ \begin{array}{l} f : (0,T) \to X; \ f \text{ measurable and} \\ \|f(t)\|_X^p \text{ is Lebesgue integrable over } (0,T) \text{ for } p \in [1,\infty) \\ \text{and } ess \sup_{t \in (0,T)} \|f(t)\|_X < \infty \text{ for } p = \infty \end{array} \right\},$$

$$W^{m,p}([0,T];X) = \{f \in \mathcal{D}'(0,T;X); \ \frac{d^j f}{dx_j} \in L^p(0,T;X), \ j = 1,\ldots,m\},$$

where $\mathcal{D}'(0,T;X)$ is the space of all continuous operators from $\mathcal{D}(0,T)$ to X. These spaces are endowed with the norms

$$\|f\|_{L^p(0,T;X)} = \left(\int_0^T \|f(t)\|_X^p \, dt \right)^{1/p},$$

$$\|f\|_{L^\infty(0,T;X)} = ess \sup_{t \in (0,T)} \|f(t)\|_X,$$

$$\|f\|_{W^{m,p}([0,T];X)} = \left(\sum_{j=1}^m \left\| \frac{d^j f}{dx^j} \right\|_{L^p(0,T;X)}^p \right)^{1/p}, \quad 1 \le p < \infty,$$

$$\|f\|_{W^{m,\infty}([0,T];X)} = \max_{1 \le j \le m} \left\| \frac{d^j f}{dx^j} \right\|_{L^\infty(0,T;X)}^p, \quad p = \infty.$$

By $C([0,T];X)$ we denote the space of continuous functions $f : [0,T] \to X$.

For simplicity, throughout the book we shall denote by $(\cdot,\cdot)$ and $\|\cdot\|$ the scalar product and the norm in $L^2(\Omega)$, respectively.

For not overloading the notation, sometimes we do not indicate in the integrands the function arguments which are the integration variables.

1.1 Well-Posedness for the Cauchy Problem with Fast Diffusion

The first section is devoted to the study of a Cauchy problem for a fast diffusion equation with transport written for the unknown function $y(t,x)$, in which the degeneracy is induced by the vanishing of the time derivative coefficient $u(x)$, on a subset of nonzero measure of the space domain. The equation is accompanied by Dirichlet boundary conditions and an initial condition set for the function $u(x)y(t,x)$.

The problem to be studied is

$$\frac{\partial(u(x)y)}{\partial t} - \Delta\beta^*(y) + \nabla \cdot K_0(x,y) \ni f \quad \text{in } Q := (0,T) \times \Omega,$$

$$y(t,x) = 0 \quad \text{on } \Sigma := (0,T) \times \Gamma, \quad (1.1)$$

$$(u(x)y(t,x))|_{t=0} = \theta_0(x) \quad \text{in } \Omega.$$

1.1.1 Hypotheses for the Parabolic–Elliptic Case

Let ρ, y_s and β_s^* be given positive constants.

In this section $\beta^* : (-\infty, y_s] \to \mathbb{R}$ is a multivalued function defined as

$$\beta^*(r) := \begin{cases} \int_0^r \beta(\xi)d\xi, & r < y_s, \\ [\beta_s^*, +\infty), & r = y_s, \end{cases} \quad (1.2)$$

where $\beta : (-\infty, y_s) \to (\rho, +\infty)$ is assumed of class $C^1(-\infty, y_s)$ and monotonically increasing on $[0, y_s)$. We also make the hypothesis that it has the behavior

$$\beta(r) \geq \gamma_\beta \left| r \right|^m + \rho, \text{ for } r \leq 0, \tag{1.3}$$

and the blow up property

$$\lim_{r \nearrow y_s} \beta(r) = +\infty, \tag{1.4}$$

such that

$$\lim_{r \nearrow y_s} \int_0^r \beta(r) = \beta_s^*. \tag{1.5}$$

The blow up property (1.4) together with (1.5) account for the fast diffusion character of the first equation in (1.1). In (1.3) $\gamma_\beta \geq 0$ and $m \geq 0$. For the sake of simplicity we can take in the diffusion nondegenerate case $\gamma_\beta = 0$ and set

$$\beta(r) = \rho > 0, \text{ for any } r \leq 0, \tag{1.6}$$

without losing the generality. In fact in the nondegenerate diffusion case the requirement is $\beta(r) \geq \rho > 0$. The more general form (1.3) can be treated in the same way. Consequently, β^* gets the properties

$$\left(\zeta - \bar{\zeta} \right) \left(r - \bar{r} \right) \geq \rho(r - \bar{r})^2, \ \forall r, \bar{r} \in (-\infty, y_s], \ \zeta \in \beta^*(r), \ \zeta \in \beta^*(\bar{r}), \tag{1.7}$$

$$\lim_{r \to -\infty} \beta^*(r) = -\infty, \tag{1.8}$$

$$\lim_{r \nearrow y_s} \beta^*(r) = \beta_s^*. \tag{1.9}$$

The definition of the weak solution which we give a little later will specify the exact meaning of the boundary value problem (1.1).

The function u is considered smooth enough, nonnegative and bounded by the upper bound u_M, that can be taken any positive constant. Hence we assume

$$u \in W^{1,\infty}(\Omega), \ 0 \leq u(x) \leq u_M \text{ for any } x \in \Omega, \tag{1.10}$$

revealing the degeneration of the equation at the points where u is zero. To be more specific we assume that

$$u(x) = 0 \text{ on } \overline{\Omega_0}, \ u(x) > 0 \text{ on } \Omega_u = \Omega \backslash \overline{\Omega_0}, \tag{1.11}$$

where Ω_0 is a fixed open bounded subset of Ω with $\text{meas}(\Omega_0) > 0$ and $\overline{\Omega_0}$ is strictly contained in Ω, see Fig. 1.1. The common boundary of Ω_0 and Ω_u is denoted $\partial\Omega_0$ and is assumed to be regular enough.

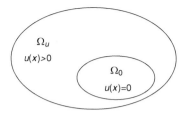

Fig. 1.1 Geometry of the problem

We also specify that the domain where u vanishes can be formed by a union of a finite number of subsets Ω_0 with the properties specified before, but we shall present the theory for only one subset.

Finally, the vector $K_0 : \Omega \times (-\infty, y_s]$ is assumed of the form

$$
K_0(x, y) = \begin{cases} a(x)K(y), & x \in \Omega_u, \\ a(x), & x \in \Omega_0, \end{cases}
$$

where $a(x) = (a_j(x))_{j=1,\dots,N}$,

$$
a_j \in W^{1,\infty}(\Omega), \ a_j(x) = 0 \text{ in } \overline{\Omega_0}, \ |a_j(x)| \leq a_j^M, \text{ for } x \in \overline{\Omega}, \qquad (1.12)
$$

and $K : (-\infty, y_s] \to \mathbb{R}$ is Lipschitz continuous, i.e., there exists $M_K > 0$ such that

$$
|K(r) - K(\overline{r})| \leq M_K \, |r - \overline{r}|, \text{ for any } r, \overline{r} \in (-\infty, y_s]. \qquad (1.13)
$$

Moreover, we assume that K is bounded

$$
|K(r)| \leq K_s, \text{ for any } r \in \mathbb{R}. \qquad (1.14)
$$

The term $\nabla \cdot K_0(x, y)$ includes both a nonlinear advection term with the velocity $a(x)K'(y)$ and a nonlinear decay or source term with the rate $\nabla \cdot a$.

1.1.2   *Functional Framework*

We begin by establishing some notation and giving a few definitions.

Let us consider the Hilbert space $V = H_0^1(\Omega)$ with the usual Hilbertian norm

$$
\|v\|_V = \left(\int_\Omega |\nabla v(x)|^2 \, dx \right)^{1/2},
$$

and its dual $V' = H^{-1}(\Omega)$.

The dual V' will be endowed with the scalar product

$$(y, \overline{y})_{V'} := \langle y, \psi \rangle_{V', V}, \qquad (1.15)$$

where $\psi \in V$ is the solution to the elliptic problem

$$A_0 \psi = \overline{y}, \qquad (1.16)$$

with $A_0 : V \to V'$ defined by

$$\langle A_0 v, \phi \rangle_{V', V} := \int_\Omega \nabla v \cdot \nabla \phi \, dx, \text{ for any } \phi \in V. \qquad (1.17)$$

The notation $\langle y, \psi \rangle_{V', V}$ represents the pairing between V' and V and it reduces to the scalar product in $L^2(\Omega)$ if $y \in L^2(\Omega)$.

It is well known that $A_0 = -\Delta$ with Dirichlet boundary conditions is the canonical isomorphism between $H_0^1(\Omega)$ and $H^{-1}(\Omega)$. Moreover, it is isometric because

$$\|y\|_{V'} = \|\psi\|_V. \qquad (1.18)$$

Indeed, by (1.15) and (1.16) we get

$$\|y\|_{V'}^2 = \langle y, \psi \rangle_{V', V} = \langle A_0 \psi, \psi \rangle_{V', V} = \|\psi\|_V^2,$$

where $\psi = A_0^{-1} y$.

We recall now the Poincaré inequality (see e.g., [30], pp. 290). Let Ω be a bounded domain in $\mathbb{R}^N$ with a sufficiently smooth boundary. For each $y \in H_0^1(\Omega)$ we have

$$\|y\| \leq c_P \|y\|_{H_0^1(\Omega)} \qquad (1.19)$$

with c_P depending only on Ω and the dimension N.

We also recall that if $\theta \in L^2(\Omega)$ we have

$$\|\theta\|_{V'} \leq c_P \|\theta\|. \qquad (1.20)$$

Indeed, by (1.15) and (1.18)

$$\|\theta\|_{V'}^2 = \langle \theta, \psi \rangle_{V', V} = \int_\Omega \theta \psi \, dx \leq \|\theta\| \, \|\psi\| \leq c_P \|\theta\| \, \|\psi\|_V = c_P \|\theta\| \, \|\theta\|_{V'}.$$

For $\theta(t) \in V'$, we denote by $\frac{d\theta}{dt}(t)$ the strong derivative of $\theta(t)$ in V', i.e.,

$$\frac{d\theta}{dt}(t) = \lim_{\varepsilon \to 0} \frac{\theta(t + \varepsilon) - \theta(t)}{\varepsilon} \text{ in } V'.$$

Finally, we specify that $u \in W^{1,\infty}(\Omega)$ is a multiplicator in V'. Let $\theta \in V'$. Noticing that $u\psi \in V$ for $\psi \in V$, we define

$$\langle u\theta, \psi \rangle_{V',V} := \langle \theta, u\psi \rangle_{V',V}, \text{ for any } \psi \in V,$$

and see by (1.15) that $u\theta$ is well defined since

$$\|u\theta\|^2_{V'} = \langle u\theta, \psi \rangle_{V',V} = \langle \theta, u\psi \rangle_{V',V} \leq \|\theta\|_{V'} \|u\psi\|_V \leq C \|\psi\|_V = C \|u\theta\|_{V'},$$

where $A_0\psi = u\theta$ and C includes the norm $\|u\|_{1,\infty} := \|u\|_{W^{1,\infty}(\Omega)}$.

Problem (1.1) will be approached under the following hypotheses for f and the initial datum:

$$f \in L^2(0, T; V'), \tag{1.21}$$

$$\theta_0 \in L^2(\Omega), \ \theta_0 = 0 \text{ a.e. on } \Omega_0,$$

$$\theta_0 \geq 0 \text{ a.e. on } \Omega_u, \ \frac{\theta_0}{u} \in L^2(\Omega_u), \ \frac{\theta_0}{u} \leq y_s, \text{ a.e. } x \in \Omega_u. \tag{1.22}$$

We recall that $\Omega_u = \Omega \backslash \overline{\Omega_0}$ and it is an open subset of Ω. The non-negativeness assumed for θ_0 is in agreement with the physical interpretation of θ_0, that of a density (in general) or a temperature. From the mathematical point of view it does not diminish the generality.

We give now the definition of a *weak solution* to (1.1).

Definition 1.1. Let (1.21) and (1.22) hold. We call a *weak solution* to (1.1) a pair (y, ζ),

$$y \in L^2(0, T; V),$$

$$\zeta \in L^2(0, T; V), \ \zeta(t, x) \in \beta^*(y(t, x)) \text{ a.e. } (t, x) \in Q, \tag{1.23}$$

$$uy \in C([0, T]; L^2(\Omega)) \cap W^{1,2}([0, T]; V'),$$

which satisfies

$$\left\langle \frac{d(uy)}{dt}(t), \psi \right\rangle_{V',V} + \int_\Omega (\nabla\zeta(t) - K_0(x, y(t))) \cdot \nabla\psi dx$$

$$= \langle f(t), \psi \rangle_{V',V}, \text{ a.e. } t \in (0, T), \text{ for any } \psi \in V, \tag{1.24}$$

the initial condition $(uy(t))|_{t=0} = \theta_0$ and the boundedness condition

$$y(t, x) \leq y_s \text{ a.e. } (t, x) \in Q. \tag{1.25}$$

It is easy to see that an equivalent form to (1.24), which will be used many times in this book is

$$\int_0^T \left\langle \frac{d(uy)}{dt}(t), \phi(t) \right\rangle_{V',V} dt + \int_Q (\nabla\zeta - K_0(x,y)) \cdot \nabla\phi dx dt$$

$$= \int_0^T \langle f(t), \phi(t) \rangle_{V',V} \, dt, \text{ for any } \phi \in L^2(0,T;V). \quad (1.26)$$

A proof of the equivalence between (1.24) and (1.26) can be found in [84], pp. 81.

We also specify that a weak solution is a solution in the sense of distributions to (1.1). Indeed if we take $\phi \in C_0^\infty(Q)$ in (1.26) we get after some computations involving Green's and Ostrogradski's formulae (see [13], pp. 13) that

$$\int_Q \left(\frac{\partial(uy)}{\partial t} - \Delta\zeta + \nabla \cdot K_0(x,y) - f \right) \phi dx dt = 0, \ \forall \phi \in C_0^\infty(Q),$$

which means that

$$\frac{\partial(uy)}{\partial t} - \Delta\zeta + \nabla \cdot K_0(x,y) - f = 0 \text{ in } \mathcal{D}'(Q).$$

The boundary condition on Σ is immediately implied by the fact that the solution $y(t) \in V = H_0^1(\Omega)$ a.e. $t \in (0,T)$.

Now we pass to the abstract writing of our problem. We set

$$D(A) := \{ y \in L^2(\Omega); \ \exists \zeta \in V, \ \zeta(x) \in \beta^*(y(x)) \text{ a.e. } x \in \Omega \}$$

and introduce the multivalued operator $A : D(A) \subset V' \to V'$ by

$$\langle Ay, \psi \rangle_{V',V} := \int_\Omega (\nabla\zeta - K_0(x,y)) \cdot \nabla\psi dx, \ \forall\psi \in V, \text{ for some } \zeta \in \beta^*(y).$$

With all these considerations we write the abstract evolution problem

$$\frac{d(uy)}{dt}(t) + Ay(t) \ni f(t), \text{ a.e. } t \in (0,T),$$

$$(uy(t))|_{t=0} = \theta_0. \quad (1.27)$$

We consider now the multiplication operator

$$M : D(A) \to L^2(\Omega), \ My := uy, \quad (1.28)$$

whose inverse M^{-1} is multivalued. Denoting

$$\theta(t, x) := u(x)y(t, x) \tag{1.29}$$

(and formally writing $y = M^{-1}\theta = \frac{\theta}{u}$) we can rewrite (1.27) in terms of θ as

$$\frac{d\theta}{dt}(t) + B\theta(t) \ni f(t), \quad \text{a.e. } t \in (0, T),$$

$$\theta(0) = \theta_0, \tag{1.30}$$

where $B = AM^{-1}$ and

$$D(B) := \left\{ \theta \in L^2(\Omega); \ \frac{\theta}{u} \in L^2(\Omega), \ \exists \zeta \in V, \ \zeta(x) \in \beta^*\left(\frac{\theta}{u}(x)\right) \text{ a.e. } x \right\}.$$

We see that $\theta \in D(B)$ implies $\theta \in L^2(\Omega)$ and $y = \frac{\theta}{u} \in D(A)$. Conversely, if $y = \frac{\theta}{u} \in D(A)$ it follows that $\theta = uy \in D(B)$.

Besides the notion of weak solution previously given we recall the concepts of strong and mild solutions (see e.g., [11, 29]). Let H be a Hilbert space and let us consider the problem

$$\frac{dz}{dt}(t) + Az(t) \ni f(t) \quad \text{a.e. } t \in (0, T),$$

$$z(0) = z_0, \tag{1.31}$$

where $A : D(A) \subset H \to H$ is a nonlinear time-independent and possibly multivalued operator. Let $f \in L^1(0, T; H)$ be given, and $z_0 \in D(A)$.

A function $z \in C([0, T]; H)$ is said to be a *strong solution* to the Cauchy problem (1.31) if z is absolutely continuous on any compact subinterval of $(0, T)$, satisfies (1.31) a.e. $t \in (0, T)$, $z(0) = z_0$ and $z(t) \in D(A)$ a.e. $t \in (0, T)$.

We remind that the absolute continuity on any compact subinterval of $(0, T)$ implies the a.e. differentiability on $(0, T)$, because H is a Hilbert space (generally this is true for a reflexive Banach space). Hence it is clear that a strong solution $z \in W^{1,1}([a, b]; H)$, for all $0 < a < b < T$.

In literature by a *mild solution* to (1.31) it is meant a continuous function which is the uniform limit of solutions to a finite difference scheme corresponding to the problem (see [10, 11]). We shall detail this definition in Chap. 2.

For a later use we still define $j : \mathbb{R} \to (-\infty, +\infty]$ by

$$j(r) := \begin{cases} \int_0^r \beta^*(\xi)d\xi, & r \leq y_s, \\ +\infty, & r > y_s. \end{cases} \tag{1.32}$$

Next, we recall the concepts of lower semicontinuity (l.s.c.) and weakly lower semicontinuity and subdifferential.

Let X be a Banach space and let $\varphi : X \to [-\infty, \infty]$. The function φ is *proper* if $\varphi(x) \neq +\infty$. The function φ is *convex* if

$$\varphi(\lambda x_1 + (1 - \lambda)x_2) \leq \lambda\varphi(x_1) + (1 - \lambda)\varphi(x_2)$$

for $\lambda \in [0, 1]$ and any x_1, $x_2 \in X$.

The function φ is said *lower semicontinuous* at $x_0 \in X$ if

$$\liminf_{x \to x_0} \varphi(x) \geq \varphi(x_0).$$

If φ is l.s.c. at each point $x_0 \in X$ then it is l.s.c. on X.

A function φ is *sequentially weakly lower semicontinuous* on X if for any sequence $(x_n)_{n \geq 1}$, $x_n \in X$, such that $x_n \rightharpoonup x$ we have

$$\varphi(x) \leq \liminf_{n \to \infty} \varphi(x_n), \ \forall x \in X.$$

Let φ be a proper convex lower semicontinuous function and let $x \in X$. The set

$$\partial\varphi(x) := \{x^* \in X'; \varphi(x) - \varphi(z) \leq \langle x^*, x - z\rangle_{X',X} , \ \forall z \in X\}$$

is called the *subdifferential* of φ at x.

Lemma 1.2. *The function j is proper, convex, lower semicontinuous and*

$$\partial j(r) = \begin{cases} \beta^*(r), & r < y_s \\ [\beta_s^*, +\infty), & r = y_s \\ \varnothing, & r > y_s. \end{cases} \tag{1.33}$$

Proof. First, we notice that

$$j(r) = \int_0^r \beta^*(\xi)d\xi \geq \frac{\rho}{2}r^2, \ \forall r \leq y_s. \tag{1.34}$$

Then, for $r \leq y_s$,

$$j(r) \leq j(y_s) = \lim_{r \nearrow y_s} \int_0^r \beta^*(\xi)d\xi \leq \lim_{r \nearrow y_s} \beta_s^* r = \beta_s^* y_s, \tag{1.35}$$

so j is proper. It is also obvious that j is convex.

We show now that j is lower semicontinuous. For $r < y_s$ the function j is continuous, so we have only to study what happens at y_s. Let us consider a sequence $(r_n)_{n \geq 1} \subset \mathbb{R}$, $r_n \leq y_s$, such that $r_n \to y_s$ and write

$$j(r_n) = \int_0^{r_n} \beta^*(\xi)d\xi = \int_0^{y_s} \chi_n(\xi)\beta^*(\xi)d\xi$$

where

$$\chi_n(\xi) = \begin{cases} 1 \text{ if } 0 \leq \xi \leq r_n, \\ 0 \text{ if } r_n < \xi \leq y_s. \end{cases}$$

We have $\chi_n(\xi)\beta^*(\xi) \geq 0$ and $\chi_n(\xi)\beta^*(\xi) \to \beta^*(\xi)$ a.e. on $(0, y_s)$ as $n \to \infty$. Using Fatou's lemma (see e.g., [13], pp. 3) we have

$$\liminf_{n\to\infty} j(r_n) = \liminf_{n\to\infty} \int_0^{y_s} \chi_n(\xi)\beta^*(\xi)d\xi \geq \int_0^{y_s} \beta^*(\xi)d\xi = j(y_s).$$

Finally we have to prove that $\beta^* = \partial j$. We begin with the inclusion $\beta^* \subset \partial j$. We have to prove that if $v \in \beta^*(r)$ then $v \in \partial j(r)$, for any $r \leq y_s$, i.e.,

$$j(r) - j(y) \leq v(r - y), \quad \text{for any } y \in \mathbb{R} \text{ and } r \leq y_s.$$

This inequality is obvious for $r < y_s$ and $y < y_s$ and for $r = y = y_s$. Let $r = y_s$ and $y < y_s$. Then we have

$$j(y_s) - j(y) = \int_y^{y_s} \beta^*(\xi)d\xi = \lim_{r \nearrow y_s} \int_y^r \beta^*(\xi)d\xi \leq \beta_s^*(y_s - y) \leq v_s(y_s - y),$$

where $v_s \in [\beta_s^*, +\infty) = \beta^*(y_s)$. If $r < y_s$ and $y = y_s$, we have

$$j(r) - j(y_s) = -\int_r^{y_s} \beta^*(\xi)d\xi$$

and this comes back to the previous situation. If $r = y_s$ and $y > y_s$, then $j(y) = +\infty$ and the inequality is verified.

Now we notice that the function β^* is maximal monotone on $\mathbb{R}$. Indeed, the range $R(I + \beta^*) = \mathbb{R}$, this being implied by the observation that the equation $r + \beta^*(r) = g \in \mathbb{R}$ has a unique solution in $(-\infty, y_s]$. In conclusion, β^* is maximal and satisfies the inclusion $\beta^* \subset \partial j$, hence it should coincide with ∂j. So, we have proved (1.33) as claimed. □

1.1.3 Approximating Problem

The approach of the Cauchy problem (1.27), or equivalently (1.30) is based on some preliminary results. Since A is multivalued due to both M^{-1} and β^* we introduce an approximating problem by regularizing both of them. In this subsection we shall study the approximating problem while in the next subsection we shall prove that it converges in some sense to (1.27).

Thus, let ε be positive and replace u by

$$u_\varepsilon(x) := u(x) + \varepsilon,$$

and β^* by a regular single-valued function $\beta_\varepsilon^* : \mathbb{R} \to \mathbb{R}$. This can be defined as a regularization of β^* using mollifiers, or for convenience it can be taken of the form

$$\beta_\varepsilon^*(r) := \begin{cases} \beta^*(r), & r < y_s - \varepsilon \\ \beta^*(y_s - \varepsilon) + \frac{\beta_s^* - \beta^*(y_s - \varepsilon)}{\varepsilon}[r - (y_s - \varepsilon)], & r \geq y_s - \varepsilon. \end{cases} \tag{1.36}$$

The function β_ε^* is differentiable and has the derivative denoted β_ε bounded on $\mathbb{R}$, for each ε positive. Also, β_ε^* is monotonically increasing on $\mathbb{R}$,

$$(\beta_\varepsilon^*(r) - \beta_\varepsilon^*(\overline{r}))\,(r - \overline{r}) \geq \rho(r - \overline{r})^2, \quad \text{for } r, \overline{r} \in \mathbb{R}, \tag{1.37}$$

and

$$\lim_{r \to -\infty} \beta_\varepsilon^*(r) = -\infty, \quad \lim_{r \to +\infty} \beta_\varepsilon^*(r) = +\infty.$$

The function K is extended for $r \geq y_s$ by its value $K(y_s) \leq K_s$, but for the sake of simplicity we denote this extension still by K. Consequently, $K_0(x, r) = a(x)K(r)$ will extend K_0 by $a(x)K(y_s)$ for $r \geq y_s$.

Then we define the single-valued operator $A_\varepsilon : D(A_\varepsilon) \subset V' \to V'$, where

$$D(A_\varepsilon) := \{y \in L^2(\Omega);\ \beta_\varepsilon^*(y) \in V\},$$

$$\langle A_\varepsilon y, \psi \rangle_{V',V} := \int_\Omega (\nabla \beta_\varepsilon^*(y) - K_0(x, y)) \cdot \nabla \psi dx, \quad \text{for any } \psi \in V, \tag{1.38}$$

and we introduce the approximating Cauchy problem

$$\frac{d(u_\varepsilon y_\varepsilon)}{dt}(t) + A_\varepsilon y_\varepsilon(t) = f(t), \quad \text{a.e. } t \in (0, T),$$

$$u_\varepsilon y_\varepsilon(0) = \theta_0. \tag{1.39}$$

Denoting now $\theta_\varepsilon := u_\varepsilon y_\varepsilon$ we can write the equivalent approximating Cauchy problem in terms of θ_ε,

$$\frac{d\theta_\varepsilon}{dt}(t) + B_\varepsilon \theta_\varepsilon(t) = f(t), \quad \text{a.e. } t \in (0, T),$$

$$\theta_\varepsilon(0) = \theta_0. \tag{1.40}$$

The operator $B_\varepsilon : D(B_\varepsilon) \subset V' \to V'$ is single-valued, has the domain

$$D(B_\varepsilon) := \left\{ v \in L^2(\Omega); \ \beta_\varepsilon^* \left(\frac{v}{u_\varepsilon} \right) \in V \right\}$$

and is defined by

$$\langle B_\varepsilon v, \psi \rangle_{V',V} := \int_\Omega \left(\nabla \beta_\varepsilon^* \left(\frac{v}{u_\varepsilon} \right) - K_0 \left(x, \frac{v}{u_\varepsilon} \right) \right) \cdot \nabla \psi dx, \ \text{for any } \psi \in V.$$

$$(1.41)$$

In fact we note that $B_\varepsilon v = A_\varepsilon \left(\frac{v}{u_\varepsilon} \right)$ and $v \in D(B_\varepsilon)$ is equivalent to $\frac{v}{u_\varepsilon} \in D(A_\varepsilon)$.

Also, it is easily seen that $D(B_\varepsilon) = V$. Indeed, if $v \in D(B_\varepsilon)$ it follows that $\frac{v}{u_\varepsilon} \in V$ by the fact that the inverse of β_ε^* is Lipschitz, and from here we get that $v \in V$, since $u_\varepsilon \in W^{1,\infty}(\Omega)$. Conversely, $v \in V$ implies $\frac{v}{u_\varepsilon} \in V$ and taking into account that the derivative of β_ε^* is bounded for each $\varepsilon > 0$ we obtain that $\beta_\varepsilon^* \left(\frac{v}{u_\varepsilon} \right) \in V$. We recall that $u_\varepsilon = u + \varepsilon \in W^{1,\infty}(\Omega)$.

Definition 1.3. Let (1.21) and (1.22) hold. We call a *strong solution* to (1.40) a function

$$\theta_\varepsilon \in C([0,T]; L^2(\Omega)) \cap W^{1,2}([0,T]; V'), \ \beta_\varepsilon^* \left(\frac{\theta_\varepsilon}{u_\varepsilon} \right) \in L^2(0,T; V),$$

that satisfies (1.40), which can be still written

$$\left\langle \frac{d\theta_\varepsilon}{dt}(t), \psi \right\rangle_{V',V} + \int_\Omega \left(\nabla \beta_\varepsilon^* \left(\frac{\theta_\varepsilon}{u_\varepsilon} \right) - K_0 \left(x, \frac{\theta_\varepsilon}{u_\varepsilon} \right) \right) \cdot \nabla \psi dx$$

$$= \langle f(t), \psi \rangle_{V',V}, \ \text{a.e. } t \in (0,T), \ \text{for any } \psi \in V \qquad (1.42)$$

and $\theta_\varepsilon(0) = \theta_0$.

Since by $\theta_\varepsilon := u_\varepsilon y_\varepsilon$, problems (1.40) and (1.39) are equivalent, it means that if θ_ε is a solution to (1.42) then y_ε is a solution to (1.39) and belongs to the same spaces as θ_ε.

An equivalent form to (1.42) can be written as

$$\int_0^T \left\langle \frac{d(u_\varepsilon y_\varepsilon)}{dt}(t), \phi(t) \right\rangle_{V',V} dt + \int_Q (\nabla \beta_\varepsilon^*(y_\varepsilon) - K_0(x, y_\varepsilon)) \cdot \nabla \phi dx dt$$

$$= \int_0^T \langle f(t), \phi(t) \rangle_{V',V}, \ \text{for any } \phi \in L^2(0,T; V). \qquad (1.43)$$

1.1.4   Existence for the Approximating Problem

First we shall prove that, for each $\varepsilon > 0$, (1.40) has a unique solution θ_ε and consequently, (1.39) has a unique solution in their appropriate functional spaces. The proof is essentially based on the quasi m-accretivity of the operator B_ε on V'. Because we are working in Hilbert spaces, we recall the celebrated theorem of Minty (see [79], or [14], pp. 34), by which the notion of a maximal monotone operator is equivalent with that of m-accretive operator.

We say that B_ε is *quasi m-accretive* on V' if $\lambda I + B_\varepsilon$ is *monotone*,

$$((\lambda I + B_\varepsilon)\theta - (\lambda I + B_\varepsilon)\bar\theta, \theta - \bar\theta)_{V'} \geq 0, \ \forall \theta, \bar\theta \in D(B_\varepsilon),$$

and surjective,

$$R(\lambda I + B_\varepsilon) = V',$$

for all $\lambda > \lambda_0$.

Lemma 1.4. *The operator B_ε is quasi m-accretive on V'.*

Proof. Let $\theta, \bar\theta \in D(B_\varepsilon)$. We compute

$$\left(B_\varepsilon\theta - B_\varepsilon\bar\theta, \theta - \bar\theta\right)_{V'} = \int_\Omega \nabla\left(\beta_\varepsilon^*\left(\frac{\theta}{u_\varepsilon}\right) - \beta_\varepsilon^*\left(\frac{\bar\theta}{u_\varepsilon}\right)\right) \cdot \nabla\psi dx$$

$$- \int_\Omega \left(K_0\left(x, \frac{\theta}{u_\varepsilon}\right) - K_0\left(x, \frac{\bar\theta}{u_\varepsilon}\right)\right) \cdot \nabla\psi dx$$

where $\psi \in V$ is the solution to $A_0\psi = \theta - \bar\theta$. Recalling (1.12)–(1.13) and that $\varepsilon \leq u_\varepsilon(x) \leq u_M + \varepsilon$ we have

$$\int_\Omega \left(K_0\left(x, \frac{\theta}{u_\varepsilon}\right) - K_0\left(x, \frac{\bar\theta}{u_\varepsilon}\right)\right) \cdot \nabla\psi dx$$

$$\leq \sum_{j=1}^N \int_{\Omega_u} M_K |a_j(x)| \left|\frac{\theta}{u_\varepsilon} - \frac{\bar\theta}{u_\varepsilon}\right| \left|\frac{\partial\psi}{\partial x_j}\right| dx$$

$$\leq \sum_{j=1}^N M_K a_j^M \left\|\frac{\theta - \bar\theta}{u_\varepsilon}\right\|_{L^2(\Omega_u)} \|\nabla\psi\|_{L^2(\Omega_u)}$$

$$\leq \frac{\overline{M}}{\varepsilon} \|\theta - \bar\theta\| \|\psi\|_V = \frac{\overline{M}}{\varepsilon} \|\theta - \bar\theta\| \|\theta - \bar\theta\|_{V'}, \tag{1.44}$$

where we have denoted $\overline{M} = M_K \sum\limits_{j=1}^{N} a_j^M$. Next, taking into account (1.37) we compute

$$((\lambda I + B_\varepsilon)\theta - (\lambda I + B_\varepsilon)\overline{\theta}, \theta - \overline{\theta})_{V'}$$

$$= \lambda \left\|\theta - \overline{\theta}\right\|_{V'}^2 + \left(B_\varepsilon\theta - B_\varepsilon\overline{\theta}, \theta - \overline{\theta}\right)_{V'}$$

$$\geq \lambda \left\|\theta - \overline{\theta}\right\|_{V'}^2 + \int_\Omega \left(\beta_\varepsilon^*\left(\frac{\theta}{u_\varepsilon}\right) - \beta_\varepsilon^*\left(\frac{\overline{\theta}}{u_\varepsilon}\right)\right)(\theta - \overline{\theta})dx$$

$$- \frac{\overline{M}}{\varepsilon}\left\|\theta - \overline{\theta}\right\|\left\|\theta - \overline{\theta}\right\|_{V'}$$

$$\geq \lambda \left\|\theta - \overline{\theta}\right\|_{V'}^2 + \frac{\rho}{2(u_M + \varepsilon)}\left\|\theta - \overline{\theta}\right\|^2 - \frac{\overline{M}^2}{2\varepsilon^2}\frac{u_M + \varepsilon}{\rho}\left\|\theta - \overline{\theta}\right\|_{V'}^2$$

$$= \left(\lambda - \frac{\overline{M}^2}{2\varepsilon^2}\frac{u_M + \varepsilon}{\rho}\right)\left\|\theta - \overline{\theta}\right\|_{V'}^2 + \frac{\rho}{2(u_M + \varepsilon)}\left\|\theta - \overline{\theta}\right\|^2, \qquad (1.45)$$

so that B_ε is quasi-monotone for $\lambda \geq \lambda_0 = \frac{\overline{M}^2(u_M+\varepsilon)}{2\rho\varepsilon^2}$. We recall that ε is positive fixed.

Next we have to prove that $R(\lambda I + B_\varepsilon) = V'$ for λ large, i.e., to show that the equation

$$\lambda\theta_\varepsilon + B_\varepsilon\theta_\varepsilon = g \qquad (1.46)$$

has a solution $\theta_\varepsilon \in D(B_\varepsilon)$ for any $g \in V'$. If we denote $\beta_\varepsilon^*\left(\frac{\theta_\varepsilon}{u_\varepsilon}\right) = \zeta \in V$, due to the fact that β_ε^* is continuous and monotonically increasing on $\mathbb{R}$ and $R(\beta_\varepsilon^*) = (-\infty, \infty)$ it follows that its inverse

$$G\zeta := u_\varepsilon(\beta_\varepsilon^*)^{-1}(\zeta) \qquad (1.47)$$

is continuous from V to $L^2(\Omega)$. Indeed, for $\zeta, \overline{\zeta} \in V$

$$\left\|G\zeta - G\overline{\zeta}\right\| = \left\|u_\varepsilon\left((\beta_\varepsilon^*)^{-1}(\zeta) - (\beta_\varepsilon^*)^{-1}(\overline{\zeta})\right)\right\| \qquad (1.48)$$

$$\leq \frac{u_M + \varepsilon}{\rho}\left\|\zeta - \overline{\zeta}\right\| \leq \frac{(u_M + \varepsilon)c_P}{\rho}\left\|\zeta - \overline{\zeta}\right\|_V,$$

where we used (1.37) and Poincaré's inequality (with the constant c_P). So, (1.46) can be rewritten as

$$\lambda G\zeta + B_0\zeta = g \qquad (1.49)$$

with $B_0 : V \to V'$ defined by

$$\langle B_0\zeta, \psi\rangle_{V',V} := \int_\Omega \left(\nabla\zeta - K_0\left(x, \frac{G\zeta}{u_\varepsilon}\right)\right) \cdot \nabla\psi dx, \ \forall\psi \in V. \qquad (1.50)$$

We shall show that $\lambda G + B_0$ is surjective. First we have

$$
\langle (\lambda G + B_0)\zeta - (\lambda G + B_0)\bar\zeta, \zeta - \bar\zeta \rangle_{V',V}
$$

$$
= \lambda \int_\Omega (G\zeta - G\bar\zeta)(\zeta - \bar\zeta)dx + \int_\Omega \left| \nabla(\zeta - \bar\zeta) \right|^2 dx
$$

$$
- \int_\Omega a(x) \left(K\left(\frac{G\zeta}{u_\varepsilon} \right) - K\left(\frac{G\bar\zeta}{u_\varepsilon} \right) \right) \cdot \nabla(\zeta - \bar\zeta)dx
$$

$$
\geq \int_\Omega \frac{\lambda \rho}{u_\varepsilon}(G\zeta - G\bar\zeta)^2 dx + \int_\Omega \left| \nabla(\zeta - \bar\zeta) \right|^2 dx
$$

$$
- \frac{\overline{M}}{\varepsilon} \left\| G\zeta - G\bar\zeta \right\| \left\| \zeta - \bar\zeta \right\|_V
$$

$$
\geq \left(\frac{\lambda \rho}{u_M + \varepsilon} - \frac{\overline{M}^2}{2\varepsilon^2} \right) \left\| G\zeta - G\bar\zeta \right\|^2 + \frac{1}{2} \left\| \zeta - \bar\zeta \right\|_V^2 ,
$$

so $\lambda G + B_0 : V \to V'$ is monotone and obviously coercive for $\lambda > \lambda_0$.
We recall that the operator $T : V \to V'$ is called *coercive* if

$$
\lim_{n \to \infty} \frac{\langle T z_n, z_n \rangle_{V',V}}{\left\| z_n \right\|_V} = +\infty
$$

for any sequence $(z_n)_{n \geq 1}$ with $\lim_{n \to \infty} \left\| z_n \right\|_V = +\infty$.

The inequality (1.48) implies also that the operator $\lambda G + B_0$ is continuous from V to V' and since it is monotone it follows that it is m-accretive. Being also coercive it is surjective (see [14], pp. 37). Therefore (1.49) has a solution meaning in fact that we have proved that (1.46) has a solution $\theta_\varepsilon \in D(B_\varepsilon)$, i.e., that B_ε is quasi m-accretive. $\qquad \square$

Next we give an intermediate result that will be used in the existence proof of the solution to the approximating problem.
First we define

$$
j_\varepsilon(r) := \int_0^r \beta_\varepsilon^*(\xi)d\xi, \ \forall r \in \mathbb{R}, \tag{1.51}
$$

and notice that $\partial j_\varepsilon(r) = \beta_\varepsilon^*(r)$, for any $r \in \mathbb{R}$.
Let

$$
\overline{K} = K_s(\mathrm{meas}(\Omega))^{1/2} \sum_{j=1}^N a_j^M .
$$

Proposition 1.5. *Let* $f \in L^2(0,T;V')$ *and* $\theta_0 \in L^2(\Omega)$. *Then problem (1.40) has a unique strong solution satisfying*

$$\int_\Omega u_\varepsilon(x) j_\varepsilon \left(\frac{\theta_\varepsilon}{u_\varepsilon}(t) \right) dx + \frac{1}{4} \int_0^t \left\| \frac{d\theta_\varepsilon}{d\tau}(\tau) \right\|_{V'}^2 d\tau + \frac{1}{4} \int_0^t \left\| \beta_\varepsilon^* \left(\frac{\theta_\varepsilon}{u_\varepsilon}(\tau) \right) \right\|_V^2 d\tau$$

$$\leq \int_\Omega u_\varepsilon(x) j_\varepsilon \left(\frac{\theta_0}{u_\varepsilon} \right) dx + \int_0^T \|f(t)\|_{V'}^2 \, dt + \overline{K}^2 T, \ t \in [0,T]. \tag{1.52}$$

Moreover,

$$\left\| \theta_\varepsilon(t) - \overline{\theta}_\varepsilon(t) \right\|_{V'}^2 + \frac{\rho}{u_M + \varepsilon} \int_0^t \left\| (\theta_\varepsilon - \overline{\theta}_\varepsilon)(\tau) \right\|^2 d\tau$$

$$\leq e^{\left(\frac{M^2}{\varepsilon^2} \frac{u_M+\varepsilon}{\rho}+1 \right) T} \left(\left\| \theta_0 - \overline{\theta}_0 \right\|_{V'}^2 + \int_0^T \left\| f(t) - \overline{f}(t) \right\|_{V'}^2 \, dt \right) \tag{1.53}$$

where θ_ε *and* $\overline{\theta}_\varepsilon$ *are two solutions to (1.40) corresponding to the pairs of data* θ_0, f *and* $\overline{\theta}_0, \overline{f}$, *respectively.*

  In addition, if $f \in W^{1,2}([0,T];L^2(\Omega))$ *and* $\theta_0 \in V$, *then*

$$\theta_\varepsilon, y_\varepsilon, \beta_\varepsilon^*(y_\varepsilon) \in L^2(0,T;H^2(\Omega)). \tag{1.54}$$

Proof. The proof is done in two steps. At the first step we take

$$\theta_0 \in D(B_\varepsilon), \ f \in W^{1,1}([0,T];V').$$

Hence the existence of a unique solution to (1.40)

$$\theta_\varepsilon \in C([0,T];V') \cap W^{1,\infty}([0,T];V') \cap L^\infty(0,T;D(B_\varepsilon)),$$

$$\beta_\varepsilon^* \left(\frac{\theta_\varepsilon}{u_\varepsilon} \right) \in L^\infty(0,T;V)$$

follows from the general theorems for evolution equations with m-accretive operators (see [14], pp. 141).

  By the properties assumed for β_ε^*, we deduce by (1.37) that its inverse is Lipschitz with the constant $\frac{1}{\rho}$, hence $\beta_\varepsilon^* \left(\frac{\theta_\varepsilon}{u_\varepsilon}(t) \right) \in D(B_\varepsilon) = H_0^1(\Omega)$ implies $\frac{\theta_\varepsilon}{u_\varepsilon}(t) \in H^1(\Omega)$, a.e. t. Since $(\beta_\varepsilon^*)^{-1}(0) = 0$ the trace of $\frac{\theta_\varepsilon}{u_\varepsilon}(t)$ (see [13], pp. 122) makes sense and vanishes on Γ. Therefore $\frac{\theta_\varepsilon}{u_\varepsilon} \in L^\infty(0,T;V)$. For proving the estimate (1.52) we test (1.40) for $\beta_\varepsilon^* \left(\frac{\theta_\varepsilon}{u_\varepsilon} \right) \in V$ and integrate over $(0, t) \times \Omega$. Taking into account the relation

$$\int_0^t \left\langle \frac{d\theta_\varepsilon}{d\tau}(\tau), \beta_\varepsilon^* \left(\frac{\theta_\varepsilon}{u_\varepsilon}(\tau) \right) \right\rangle_{V',V} d\tau$$

$$= \int_0^t \int_\Omega u_\varepsilon(x) \frac{d}{d\tau} \left(j_\varepsilon \left(\frac{\theta_\varepsilon}{u_\varepsilon}(t) \right) \right) dx d\tau$$

$$= \int_\Omega u_\varepsilon(x) j_\varepsilon \left(\frac{\theta_\varepsilon}{u_\varepsilon}(t) \right) dx - \int_\Omega u_\varepsilon(x) j_\varepsilon \left(\frac{\theta_0}{u_\varepsilon} \right) dx,$$

we obtain that

$$\int_\Omega u_\varepsilon(x) j_\varepsilon \left(\frac{\theta_\varepsilon}{u_\varepsilon}(t) \right) dx + \int_0^t \left\| \beta_\varepsilon^* \left(\frac{\theta_\varepsilon}{u_\varepsilon}(\tau) \right) \right\|_V^2 d\tau$$

$$\leq \int_\Omega u_\varepsilon(x) j_\varepsilon \left(\frac{\theta_0}{u_\varepsilon} \right) dx + \int_0^t \| f(\tau) \|_{V'} \left\| \beta_\varepsilon^* \left(\frac{\theta_\varepsilon}{u_\varepsilon}(\tau) \right) \right\|_V d\tau$$

$$- \int_0^t \int_\Omega K_0 \left(x, \frac{\theta_\varepsilon}{u_\varepsilon}(\tau) \right) \cdot \nabla \beta_\varepsilon^* \left(\frac{\theta_\varepsilon}{u_\varepsilon}(\tau) \right) dx d\tau.$$

From there, using (1.14) we get

$$\int_\Omega u_\varepsilon(x) j_\varepsilon \left(\frac{\theta_\varepsilon}{u_\varepsilon}(t) \right) dx + \frac{1}{2} \int_0^t \left\| \beta_\varepsilon^* \left(\frac{\theta_\varepsilon}{u_\varepsilon}(\tau) \right) \right\|_V^2 d\tau$$

$$\leq \int_\Omega u_\varepsilon(x) j_\varepsilon \left(\frac{\theta_0}{u_\varepsilon} \right) dx + \int_0^T \| f(t) \|_{V'}^2 dt + \overline{K}^2 T, \text{ for } t \in [0, T]. \quad (1.55)$$

Next, we multiply (1.40) scalarly in V' by $\frac{d\theta_\varepsilon}{dt}$ and integrate over $(0, t)$. By similar computations based on the definition of the scalar product in V', we get

$$\frac{1}{2} \int_0^t \left\| \frac{d\theta_\varepsilon}{d\tau}(\tau) \right\|_{V'}^2 d\tau + \int_\Omega u_\varepsilon(x) j_\varepsilon \left(\frac{\theta_\varepsilon}{u_\varepsilon}(t) \right) dx \qquad (1.56)$$

$$\leq \int_\Omega u_\varepsilon(x) j_\varepsilon \left(\frac{\theta_0}{u_\varepsilon} \right) dx + \int_0^T \| f(t) \|_{V'}^2 dt + \overline{K}^2 T.$$

Adding the previous two inequalities we obtain (1.52).

In the second step we take

$$\theta_0 \in L^2(\Omega) = \overline{D(B_\varepsilon)}, \ f \in L^2(0, T; V').$$

Since $W^{1,1}([0, T]; V')$ is dense in $L^2(0, T; V')$ and $\overline{D(B_\varepsilon)} = V$ is dense in $L^2(\Omega)$ we can take the sequences $(f_n)_{n\geq 1} \subset W^{1,1}([0, T]; V')$ and $(\theta_0^n)_{n\geq 1} \subset D(B_\varepsilon)$ such that

$$f_n \to f \text{ strongly in } L^2(0, T; V'),$$

$$\theta_0^n \to \theta_0 \text{ strongly in } L^2(\Omega) \text{ as } n \to \infty.$$

Then, for each $\varepsilon > 0$, the problem

$$\frac{d\theta_\varepsilon^n}{dt}(t) + B_\varepsilon \theta_\varepsilon^n(t) = f_n(t), \text{ a.e. } t \in (0, T), \tag{1.57}$$

$$\theta_\varepsilon^n(0) = \theta_0^n$$

has, according to the first step, a unique solution θ_ε^n satisfying the estimate (1.52), namely,

$$\int_\Omega u_\varepsilon(x) j_\varepsilon \left(\frac{\theta_\varepsilon^n}{u_\varepsilon}(t)\right) dx + \frac{1}{4} \int_0^t \left\| \frac{d\theta_\varepsilon^n}{d\tau}(\tau) \right\|_{V'}^2 d\tau + \frac{1}{4} \int_0^t \left\| \beta_\varepsilon^* \left(\frac{\theta_\varepsilon^n}{u_\varepsilon}(\tau)\right) \right\|_V^2 d\tau$$

$$\leq \int_\Omega u_\varepsilon(x) j_\varepsilon \left(\frac{\theta_0^n}{u_\varepsilon}\right) dx + \int_0^T \|f_n(t)\|_{V'}^2 dt + \overline{K}^2 T, \tag{1.58}$$

for any $t \in [0, T]$. We stress that ε is fixed.

We notice that j_ε is Lipschitz and by the definition of β_ε^* and j_ε we have

$$\int_\Omega u_\varepsilon(x) j_\varepsilon \left(\frac{\theta_0^n}{u_\varepsilon}\right) dx \leq (u_M + \varepsilon) \frac{\beta_s^* - \beta^*(y_s - \varepsilon)}{2\varepsilon} \left\| \frac{\theta_0^n}{u_\varepsilon} \right\|^2, \tag{1.59}$$

whence

$$\int_\Omega u_\varepsilon(x) j_\varepsilon \left(\frac{\theta_\varepsilon^n}{u_\varepsilon}(t)\right) dx + \frac{1}{4} \int_0^t \left\| \frac{d\theta_\varepsilon^n}{d\tau}(\tau) \right\|_{V'}^2 d\tau + \frac{1}{4} \int_0^t \left\| \beta_\varepsilon^* \left(\frac{\theta_\varepsilon^n}{u_\varepsilon}(\tau)\right) \right\|_V^2 d\tau$$

$$\leq (u_M + \varepsilon) \frac{\beta_s^* - \beta^*(y_s - \varepsilon)}{2\varepsilon} \left\| \frac{\theta_0^n}{u_\varepsilon} \right\|^2 + \int_0^T \|f_n(t)\|_{V'}^2 dt + \overline{K}^2 T \tag{1.60}$$

$$\leq (u_M + \varepsilon) \frac{\beta_s^* - \beta^*(y_s - \varepsilon)}{\varepsilon} \left\| \frac{\theta_0}{u_\varepsilon} \right\|^2 + \int_0^T \|f(t)\|_{V'}^2 dt + \overline{K}^2 T + 2\varepsilon,$$

due to the strong convergence $\theta_0^n \to \theta_0$ and $f_n \to f$ as $n \to \infty$. Thus the right-hand side in (1.60) is independent of n, since ε is small, fixed, e.g. $\varepsilon \ll 1$.

Recalling (1.34), $j_\varepsilon(r) \geq \frac{\rho}{2}r^2$ for any $r \in \mathbb{R}$, we can write by (1.60) that

$$\frac{\rho}{(u_M + \varepsilon)} \|\theta_\varepsilon^n(t)\|^2$$

$$\leq (u_M+1)\frac{\beta_s^* - \beta^*(y_s - \varepsilon)}{\varepsilon} \left\|\frac{\theta_0}{u_\varepsilon}\right\|^2 + \int_0^T \|f(t)\|_{V'}^2 \, dt + \overline{K}^2 T + 2, \quad (1.61)$$

for any $t \in [0, T]$.

We deduce that $\left(\beta_\varepsilon^* \left(\frac{\theta_\varepsilon^n}{u_\varepsilon}\right)\right)_n$ lies in a bounded subset of $L^2(0, T; V)$ and $\left(\frac{d\theta_\varepsilon^n}{dt}\right)_n$ is in a bounded subset of $L^2(0, T; V')$. Therefore we can select a subsequence, denoted still by the subscript n, such that

$$\frac{d\theta_\varepsilon^n}{dt} \rightharpoonup \frac{d\theta_\varepsilon}{dt} \text{ in } L^2(0, T; V') \text{ as } n \to \infty,$$

$$\beta_\varepsilon^* \left(\frac{\theta_\varepsilon^n}{u_\varepsilon}\right) \rightharpoonup \zeta_\varepsilon \text{ in } L^2(0, T; V) \text{ as } n \to \infty.$$

The latter immediately implies that

$$\frac{\theta_\varepsilon^n}{u_\varepsilon} \rightharpoonup y_\varepsilon \text{ in } L^2(0, T; V) \text{ as } n \to \infty.$$

But $u_\varepsilon \in W^{1,\infty}(\Omega)$ and the sequence $(\theta_\varepsilon)_n = \left(u_\varepsilon \frac{\theta_\varepsilon^n}{u_\varepsilon}\right)_n$ is bounded in $L^2(0, T; V)$ so that we get

$$\theta_\varepsilon^n \rightharpoonup \theta_\varepsilon \text{ in } L^2(0, T; V) \text{ as } n \to \infty.$$

At this point we recall the following theorem (see [7, 77]).

Theorem (Aubin–Lions). *Let X_1, X_2, X_3 be three Banach spaces, X_1 and X_3 reflexive, $X_1 \subset X_2 \subset X_3$ with dense and continuous inclusions and the inclusion $X_1 \subset X_2$ is compact. Let $(z_n)_{n \geq 1}$ be a bounded sequence in $L^{p_1}(0, T; X_1)$ such that $(\frac{dz_n}{dt})_{n \geq 1}$ is bounded in $L^{p_3}(0, T; X_3)$. Then $(z_n)_{n \geq 1}$ is compact in $L^{p_2}(0, T; X_2)$, where $1 \leq p_1, p_2, p_3 < \infty$.*

On the basis of the previous convergencies and since V is compact in $L^2(\Omega)$ it follows by the above theorem that

$$\theta_\varepsilon^n \to \theta_\varepsilon \text{ in } L^2(0, T; L^2(\Omega)) \text{ as } n \to \infty$$

and also (since $u_\varepsilon \geq \varepsilon$) that

$$\frac{\theta_\varepsilon^n}{u_\varepsilon} \to \frac{\theta_\varepsilon}{u_\varepsilon} \text{ in } L^2(0, T; L^2(\Omega)) \text{ as } n \to \infty.$$

By (1.36) we have

$$\left\| \beta_\varepsilon^* \left(\frac{\theta_\varepsilon^n}{u_\varepsilon} \right) - \beta_\varepsilon^* \left(\frac{\theta_\varepsilon}{u_\varepsilon} \right) \right\|_{L^2(Q)} = \left\| \frac{\beta_s^* - \beta^*(y_s - \varepsilon)}{\varepsilon} \left(\frac{\theta_\varepsilon^n}{u_\varepsilon} - \frac{\theta_\varepsilon}{u_\varepsilon} \right) \right\|_{L^2(Q)}$$

and deduce that

$$\beta_\varepsilon^* \left(\frac{\theta_\varepsilon^n}{u_\varepsilon} \right) \to \beta_\varepsilon^* \left(\frac{\theta_\varepsilon}{u_\varepsilon} \right) \text{ in } L^2(Q), \text{ as } n \to \infty,$$

hence $\zeta_\varepsilon = \beta_\varepsilon^* \left(\frac{\theta_\varepsilon}{u_\varepsilon} \right)$ a.e. on Q.

Moreover, since K is Lipschitz it follows that

$$K \left(\frac{\theta_\varepsilon^n}{u_\varepsilon} \right) \to K \left(\frac{\theta_\varepsilon}{u_\varepsilon} \right) \text{ in } L^2(0, T; L^2(\Omega)) \text{ as } n \to \infty.$$

Finally, the Ascoli–Arzelà theorem (see below) implies that

$$\theta_\varepsilon^n(t) \to \theta_\varepsilon(t) \text{ in } V', \text{ as } n \to \infty, \text{ uniformly in } t \in [0, T], \qquad (1.62)$$

as we further prove. First we recall this theorem.

Theorem (Ascoli–Arzelà). *Let X be a Banach space and let $\mathcal{M} \subset C([0, T]; X)$ be a family of functions such that*

(i) $\|u(t)\|_X \leq C, \forall t \in [0, T], u \in \mathcal{M}$,
(ii) $\mathcal{M}$ is equi-uniformly continuous i.e., $\forall \varepsilon, \exists \delta(\varepsilon)$ such that

$$\|u(t) - u(s)\|_X \leq \varepsilon \text{ if } |t - s| \leq \delta(\varepsilon), \forall u \in \mathcal{M},$$

(iii) For each $t \in [0, T]$ the set $\{u(t); u \in \mathcal{M}\}$ is compact in X.

Then, $\mathcal{M}$ is compact in $C([0, T]; X)$.

Indeed, the family $\mathcal{M} = (\theta_\varepsilon^n)_n \subset C([0, T]; V')$ is bounded (this follows e.g., by (1.61)) and equi-uniformly continuous. To prove this, let $\varepsilon' > 0$ and consider that $\sigma(\varepsilon')$ exists such that $|t - s| \leq \sigma(\varepsilon')$, for $0 \leq s < t \leq T$. We have

$$\|\theta_\varepsilon^n(t) - \theta_\varepsilon^n(s)\|_{V'} = \left\| \int_s^t \frac{d\theta_\varepsilon^n}{dt}(\tau) d\tau \right\|_{V'} \leq \int_s^t \left\| \frac{d\theta_\varepsilon^n}{dt}(\tau) \right\|_{V'} d\tau$$

$$\leq |t - s|^{1/2} \left\| \frac{d\theta_\varepsilon^n}{dt} \right\|_{L^2(0, T; V')} \leq \varepsilon', \text{ for } \sigma(\varepsilon') \leq \frac{\varepsilon'^2}{\gamma_0(\varepsilon)}, \forall \theta_\varepsilon^n \in \mathcal{M},$$

where $\gamma_0(\varepsilon)$ is the right-hand side in (1.60) which is independent of n. Still by (1.61) we get that the sequence $(\theta_\varepsilon^n(t))_n$ is bounded in $L^2(\Omega)$ for any $t \in [0, T]$ and since the injection of $L^2(\Omega)$ in V' is compact it follows that the sequence $(\theta_\varepsilon^n(t))_n$ is compact in V', for each $t \in [0, T]$. Hence the set $\mathcal{M}$ is compact in $C([0, T]; V')$, i.e., we have (1.62).

From here we get that $\lim\limits_{n \to \infty} \theta_\varepsilon^n(0) = \theta_\varepsilon(0)$, whence $\theta_0 = \theta_\varepsilon(0)$.

By (1.57) we have that

$$B_\varepsilon \theta_\varepsilon^n = f_n - \frac{d\theta_\varepsilon^n}{dt} \rightharpoonup f - \frac{d\theta_\varepsilon}{dt} \text{ in } L^2(0, T; V'), \text{ as } n \to \infty.$$

Since B_ε is quasi m-accretive on V', its realization on $L^2(0, T; V')$ is quasi m-accretive too, hence it is demiclosed and the previous weak convergence together with the strong convergence $\theta_\varepsilon^n \to \theta_\varepsilon$ leads to

$$B_\varepsilon \theta_\varepsilon = f - \frac{d\theta_\varepsilon}{dt} \text{ in } L^2(0, T; V'),$$

(see [14], pp.100). We recall that a subset A of $X \times X$ is called *demiclosed* if it is strongly–weakly closed in $X \times X$, i.e., $z_n \to z$, $w_n \rightharpoonup w$ where $w_n \in Az_n$ imply $w \in Az$. Thus, we have got (1.40), and proved that this problem has the solution $\theta_\varepsilon \in C([0, T], L^2(\Omega)) \cap W^{1,2}([0, T]; V') \cap L^2(0, T; V)$.

Finally, passing to limit in (1.58) as $n \to \infty$, and using the lower semicontinuity property we get (1.52), as claimed.

Consider now two problems (1.40) corresponding to the pairs of data θ_0, f and $\overline{\theta}_0, \overline{f}$. They have the solutions denoted θ_ε and $\overline{\theta}_\varepsilon$, respectively. We subtract the equations and multiply the difference by $(\theta_\varepsilon - \overline{\theta}_\varepsilon)(t)$, scalarly in V'. Then we integrate it over $(0, t)$. A few calculations on the basis of (1.45) lead us to

$$\left\| \theta_\varepsilon(t) - \overline{\theta}_\varepsilon(t) \right\|_{V'}^2 + \frac{\rho}{u_M + \varepsilon} \int_0^t \left\| \theta_\varepsilon(\tau) - \overline{\theta}_\varepsilon(\tau) \right\|^2 d\tau \leq \left\| \theta_0 - \overline{\theta}_0 \right\|_{V'}^2$$

$$+ \int_0^T \left\| f(t) - \overline{f}(t) \right\|_{V'}^2 dt + \left(\frac{\overline{M}^2(u_M + \varepsilon)}{\varepsilon^2 \rho} + 1 \right) \int_0^t \left\| (\theta_\varepsilon - \overline{\theta}_\varepsilon)(\tau) \right\|_{V'}^2 d\tau$$

which by the Gronwall's lemma implies (1.53). This also implies the uniqueness if the data are the same.

Finally, we give an idea for the proof of (1.54). Let $f \in W^{1,2}([0, T]; L^2(\Omega))$ and $\theta_0 \in V$. A rigorous computation means to replace (1.40) by a time finite difference equation, to multiply it by $\frac{\beta_\varepsilon^*(y_\varepsilon(t+h)) - \beta_\varepsilon^*(y_\varepsilon(t))}{h}$ which is in V and to integrate with respect to t. For simplicity we present a more formal computation. We multiply (1.40) by $\frac{\partial \beta_\varepsilon^*(y_\varepsilon)}{\partial t}$ and integrate over $(0, t) \times \Omega$. We get

$$\int_0^t \int_\Omega u_\varepsilon \beta_\varepsilon(y_\varepsilon) \left(\frac{dy_\varepsilon}{d\tau} \right)^2 dx d\tau + \frac{1}{2} \int_0^t \frac{d}{d\tau} \left\| \nabla \beta_\varepsilon^*(y_\varepsilon(\tau)) \right\|^2 d\tau$$

$$= \int_0^t \int_\Omega a(x) K(y_\varepsilon) \cdot \nabla \left(\frac{d\beta_\varepsilon^*(y_\varepsilon(\tau))}{d\tau} \right) dx d\tau + \int_0^t \int_\Omega f \frac{d\beta_\varepsilon^*(y_\varepsilon)}{d\tau} dx d\tau.$$

After the integration with respect to τ in the second term on the left-hand side, we obtain

$$\int_0^t \int_\Omega u_\varepsilon \beta_\varepsilon(y_\varepsilon) \left(\frac{dy_\varepsilon}{d\tau}\right)^2 dx d\tau + \frac{1}{2}\|\beta_\varepsilon^*(y_\varepsilon(t))\|_V^2 - \frac{1}{2}\|\beta_\varepsilon^*(y_\varepsilon(0))\|_V^2$$

$$= \int_\Omega a(x)K(y_\varepsilon(t)) \cdot \nabla\beta_\varepsilon^*(y_\varepsilon(t))dx - \int_\Omega a(x)K(y_\varepsilon(0)) \cdot \nabla\beta_\varepsilon^*(y_\varepsilon(0))dx$$

$$- \int_0^t \int_\Omega a(x)\frac{\partial K(y_\varepsilon)}{\partial \tau} \cdot \nabla\beta_\varepsilon^*(y_\varepsilon(\tau))dx d\tau$$

$$+ \int_\Omega f(t)\beta_\varepsilon^*(y_\varepsilon(t))dx - \int_\Omega f(0)\beta_\varepsilon^*(y_\varepsilon(0))dx - \int_0^t \int_\Omega \frac{\partial f}{\partial \tau}\beta_\varepsilon^*(y_\varepsilon)dx d\tau.$$

Next we have

$$\int_0^t \int_\Omega u_\varepsilon \beta_\varepsilon(y_\varepsilon) \left(\frac{dy_\varepsilon}{d\tau}\right)^2 d\tau dx + \frac{1}{2}\|\beta_\varepsilon^*(y_\varepsilon(t))\|_V^2$$

$$\leq C_0(\varepsilon) + \overline{M}\|y_\varepsilon(t)\|\,\|\beta_\varepsilon^*(y_\varepsilon(t))\|_V + \overline{M}\int_0^t \left\|\frac{dy_\varepsilon}{d\tau}(\tau)\right\|\,\|\beta_\varepsilon^*(y_\varepsilon(\tau))\|_V\,d\tau$$

$$+ c_P\|f(t)\|\,\|\beta_\varepsilon^*(y_\varepsilon(t))\|_V + c_P\int_0^t \left\|\frac{\partial f}{\partial \tau}(\tau)\right\|\,\|\beta_\varepsilon^*(y_\varepsilon(\tau))\|_V\,d\tau,$$

where

$$C_0(\varepsilon) = \frac{1}{2}\left\|\beta_\varepsilon^*\left(\frac{\theta_0}{u_\varepsilon}\right)\right\|_V^2 + \overline{M}\left\|\frac{\theta_0}{u_\varepsilon}\right\|\,\left\|\beta_\varepsilon^*\left(\frac{\theta_0}{u_\varepsilon}\right)\right\|_V + c_P\|f(0)\|\,\left\|\beta_\varepsilon^*\left(\frac{\theta_0}{u_\varepsilon}\right)\right\|_V.$$

$$(1.63)$$

By $\beta_\varepsilon(y_\varepsilon) \geq \rho$ and (1.52) we deduce

$$\rho\varepsilon \int_0^t \int_\Omega \left(\frac{\partial y_\varepsilon}{\partial \tau}\right)^2 dx d\tau + \frac{1}{4}\|\beta_\varepsilon^*(y_\varepsilon)\|_V^2$$

$$\leq C_0(\varepsilon) + \frac{\rho}{2}\int_0^t \int_\Omega \varepsilon\left(\frac{\partial y_\varepsilon}{\partial \tau}\right)^2 dx d\tau + \frac{1}{2}\left(\frac{\overline{M}^2}{\rho\varepsilon} + 1\right)\int_0^t \|\beta_\varepsilon^*(y_\varepsilon(\tau))\|_V^2\,d\tau$$

$$+ 2\overline{M}^2\|y_\varepsilon(t)\|^2 + 2c_P^2\|f(t)\|^2 + \frac{c_P^2}{2}\int_0^t \left\|\frac{\partial f}{\partial \tau}(\tau)\right\|^2\,d\tau,$$

whence we get $\frac{dy_\varepsilon}{dt} \in L^2(Q)$, $\beta_\varepsilon^*(y_\varepsilon) \in L^\infty(0,T;V)$ for each $\varepsilon > 0$.

We continue with some other computations based on the arguments developed in [84], Theorem 2.6, pp. 156. These are very long and technical so we do no longer provide them. We obtain an estimate of the form

$$\|\beta_\varepsilon^*(y_\varepsilon)\|^2_{W^{1,2}([0,T];L^2(\Omega))} + \|\beta_\varepsilon^*(y_\varepsilon)\|^2_{L^\infty(0,T;V)} + \|\beta_\varepsilon^*(y_\varepsilon)\|^2_{L^2(0,T;H^2(\Omega))}$$

$$\leq \gamma_1 \frac{\beta_s^* - \beta^*(y_s - \varepsilon)}{\varepsilon}$$

$$\times \left(\left\|\beta_\varepsilon^*\left(\frac{\theta_0}{u+\varepsilon}\right)\right\|^2_V + \int_\Omega j_\varepsilon\left(\frac{\theta_0}{u+\varepsilon}\right) dx + \|f(t)\|^2_{W^{1,2}([0,T];L^2(\Omega))} + 1 \right),$$

$$(1.64)$$

where γ_1 is a constant depending on the problem data. Since $\theta_0 \in V$ it follows that $\frac{\theta_0}{u+\varepsilon} \in V$ and $j_\varepsilon\left(\frac{\theta_0}{u+\varepsilon}\right) \in L^1(\Omega)$, so that by (1.64) we get that $\beta_\varepsilon^*(y_\varepsilon) \in L^2(0,T;H^2(\Omega))$. By a direct computation we also get that $a_j K(y_\varepsilon) \in L^2(0,T;H^1(\Omega))$, $j = 1,\ldots,N$.

For a later use we specify that these imply the flux continuity across a surface, i.e.,

$$(K_0(x,y_\varepsilon(t)) - \nabla\beta_\varepsilon^*(y_\varepsilon(t))) \cdot \nu \text{ is continuous across } \Gamma_c, \text{ a.e. } t \in (0,T),$$

$$(1.65)$$

where Γ_c is any surface included in Ω and ν is the outer normal to Γ_c. Indeed, since each component $\eta_i(t)$ of the flux vector belongs to $H^1(\Omega)$, a.e. t it follows that its trace on any line crossing the surface Γ_c is continuous. Therefore the normal component of the gradient is continuous across any Γ_c and in particular across $\partial\Omega_0$. □

1.1.5 Convergence of the Approximating Problem

Theorem 1.6. *Let (1.21) and (1.22) hold. Then, the Cauchy problem (1.27) has at least a weak solution (y^*, ζ).*

Proof. Let us assume (1.21) and (1.22), i.e.,

$$\theta_0 \in L^2(\Omega), \ \theta_0 = 0 \text{ a.e. on } \Omega_0,$$

$$\theta_0 \geq 0 \text{ a.e. on } \Omega_u, \ \frac{\theta_0}{u} \in L^2(\Omega_u), \ \frac{\theta_0}{u} \leq y_s, \text{ a.e. } x \in \Omega_u.$$

According to Proposition 1.5 there exists a unique solution to (1.40), with
the properties (1.52), (1.53). Then, it follows that

$$\int_\Omega j_\varepsilon\left(\frac{\theta_0}{u_\varepsilon}\right)dx = \int_{\Omega_0} j_\varepsilon\left(\frac{\theta_0}{u_\varepsilon}\right)dx + \int_{\Omega_u} j_\varepsilon\left(\frac{\theta_0}{u_\varepsilon}\right)dx = \int_{\Omega_u} j_\varepsilon\left(\frac{\theta_0}{u_\varepsilon}\right)dx$$

since $\frac{\theta_0}{u_\varepsilon} = 0$ a.e. on $\overline{\Omega_0}$. Using (1.35) and the fact that $u_\varepsilon = u + \varepsilon > u$ on
Ω_u, we still obtain

$$\int_\Omega j_\varepsilon\left(\frac{\theta_0}{u_\varepsilon}\right)dx = \int_{\Omega_u}\int_0^{\theta_0/u_\varepsilon}\beta_\varepsilon^*(r)drdx$$

$$\leq \int_{\Omega_u}\int_0^{\theta_0/u}\beta_\varepsilon^*(r)drdx \leq \beta_s^* y_s\mathrm{meas}(\Omega),$$

and so the right-hand side in (1.52) becomes essentially independent of ε,

$$\int_\Omega u_\varepsilon(x)j_\varepsilon\left(\frac{\theta_\varepsilon}{u_\varepsilon}(t)\right)dx + \int_0^t\left\|\frac{d\theta_\varepsilon}{d\tau}(\tau)\right\|_{V'}^2 d\tau + \int_0^t\left\|\beta_\varepsilon^*\left(\frac{\theta_\varepsilon(\tau)}{u_\varepsilon}\right)\right\|_V^2 d\tau$$

$$\leq 4(u_M+\varepsilon)\left(\beta_s^* y_s\mathrm{meas}(\Omega)+\int_0^T\|f(t)\|_{V'}^2\,dt + \overline{K}^2 T\right),\ t\in[0,T].\ (1.66)$$

Then, using (1.34) we get

$$\left\|\sqrt{u_\varepsilon}\frac{\theta_\varepsilon}{u_\varepsilon}(t)\right\|^2$$

$$\leq \frac{8}{\rho}(u_M + \varepsilon)\left(\beta_s^* y_s\mathrm{meas}(\Omega) + \int_0^T\|f(t)\|_{V'}^2\,dt + \overline{K}^2 T\right),\ t\in[0,T].$$

$$(1.67)$$

Next, we write again

$$\theta_\varepsilon = \left(\sqrt{u_\varepsilon}\frac{\theta_\varepsilon}{u_\varepsilon}\right)\sqrt{u_\varepsilon}$$

and obtain

$$\|\theta_\varepsilon(t)\|^2 \leq \frac{8}{\rho}(u_M+\varepsilon)^2\left(\beta_s^* y_s\mathrm{meas}(\Omega) + \int_0^T\|f(t)\|_{V'}^2\,dt + \overline{K}^2 T\right),\ t\in[0,T].$$

$$(1.68)$$

Therefore, the right-hand side terms in the estimates (1.66)–(1.68) are
bounded by constants (since ε is small, e.g., $\varepsilon \ll 1$).

1.1.5.1 Passing to the Limit as $\varepsilon \to 0$

On the basis of these estimates we can select a subsequence denoted still by the subscript ε, such that

$$\beta_\varepsilon^* \left(\frac{\theta_\varepsilon}{u_\varepsilon} \right) \rightharpoonup \zeta \text{ in } L^2(0,T;V), \text{ as } \varepsilon \to 0, \tag{1.69}$$

$$y_\varepsilon = \frac{\theta_\varepsilon}{u_\varepsilon} \rightharpoonup y \text{ in } L^2(0,T;V), \text{ as } \varepsilon \to 0, \tag{1.70}$$

$$\sqrt{u_\varepsilon} \frac{\theta_\varepsilon}{u_\varepsilon} \overset{w*}{\to} \chi \text{ in } L^\infty(0,T;L^2(\Omega)), \text{ as } \varepsilon \to 0. \tag{1.71}$$

But

$$\theta_\varepsilon = u_\varepsilon \frac{\theta_\varepsilon}{u_\varepsilon} \tag{1.72}$$

and since $u_\varepsilon \to u$ uniformly on Ω and $u \in W^{1,\infty}(\Omega)$ we have that

$$\|\theta_\varepsilon\|_{L^2(0,T;V)} \leq \text{ constant independent of } \varepsilon, \tag{1.73}$$

and so

$$\theta_\varepsilon \rightharpoonup \theta \text{ in } L^2(0,T;V), \text{ as } \varepsilon \to 0. \tag{1.74}$$

By (1.66) we still deduce that

$$\frac{d\theta_\varepsilon}{dt} \rightharpoonup \frac{d\theta}{dt} \text{ in } L^2(0,T;V'), \text{ as } \varepsilon \to 0, \tag{1.75}$$

and by (1.40) we have

$$\Delta \beta_\varepsilon^* \left(\frac{\theta_\varepsilon}{u_\varepsilon} \right) \rightharpoonup \frac{d\theta}{dt} - f \text{ in } L^2(0,T;V'), \text{ as } \varepsilon \to 0. \tag{1.76}$$

Also, by (1.70), (1.72), (1.74) and $u_\varepsilon \to u$ uniformly we deduce that

$$\theta = uy \text{ a.e. on } Q, \tag{1.77}$$

and obviously

$$\theta = 0 \text{ a.e. on } Q_0, \tag{1.78}$$

where $Q_0 := (0,T) \times \Omega_0$. Using (1.71) and (1.70) we still obtain that

$$\sqrt{u_\varepsilon} y_\varepsilon \overset{w*}{\to} \chi = \sqrt{u} y \text{ in } L^\infty(0,T;L^2(\Omega)). \tag{1.79}$$

Again, by the Ascoli–Arzelà theorem we deduce that

$$\theta_\varepsilon(t) \to \theta(t) \text{ in } V', \text{ as } \varepsilon \to 0, \text{ uniformly in } t \in [0, T]. \tag{1.80}$$

Thus,

$$\theta_0 = \lim_{\varepsilon \to 0} \theta_\varepsilon(0) = \theta(0) = (uy(t))|_{t=0}.$$

By the Aubin–Lions theorem $(\theta_\varepsilon)_\varepsilon$ is compact in $L^2(0, T; L^2(\Omega))$, i.e.,

$$\theta_\varepsilon \to \theta \text{ in } L^2(0, T; L^2(\Omega)) \text{ as } \varepsilon \to 0. \tag{1.81}$$

We set now for $\delta > 0$ arbitrarily small

$$\Omega_\delta := \{x \in \Omega; \ u(x) > \delta\}, \quad Q_\delta := (0, T) \times \Omega_\delta. \tag{1.82}$$

We recall that

$$\Omega_u := \{x \in \Omega; \ u(x) > 0\}, \quad Q_u := (0, T) \times \Omega_u \tag{1.83}$$

and notice that Ω_δ and Ω_u are open. We have

$$\frac{1}{u_\varepsilon} = \frac{1}{u + \varepsilon} < \frac{1}{\delta} \text{ on } \Omega_\delta,$$

so that, by (1.81) and (1.70) we can conclude that

$$y_\varepsilon = \frac{1}{u_\varepsilon}\theta_\varepsilon \to \frac{\theta}{u} := y \text{ in } L^2(0, T; L^2(\Omega_\delta)), \tag{1.84}$$

and a.e. in Q_δ, $\forall \delta > 0$. Still by (1.70) we have that

$$y_\varepsilon = \frac{\theta_\varepsilon}{u_\varepsilon} \rightharpoonup y \text{ in } L^2(0, T; L^2(\Omega_u)). \tag{1.85}$$

1.1.5.2 Convergence of $\beta_\varepsilon^*(y_\varepsilon)$ on Q_u

Let $(t, x) \in Q_\delta$. First, we shall prove that

$$\zeta(t, x) \in \beta^*(y(t, x)) \text{ a.e. on } Q_\delta, \tag{1.86}$$

where ζ is given by (1.69). This will be proved using the fact that j is the potential of β^*, i.e., $\beta^* = \partial j$.

To this end we establish some relations. We note that

$$j_\varepsilon(z) \to j(z), \quad \text{as } \varepsilon \to 0, \quad \text{for any } z \in \mathbb{R}. \tag{1.87}$$

This assertion is clear for $z < y_s - \varepsilon$, where $j_\varepsilon(z) \equiv j(z)$.
For $y_s - \varepsilon \leq z < y_s$ we compute

$$|j_\varepsilon(z) - j(z)| = \left| \int_{y_s - \varepsilon}^z (\beta_\varepsilon^*(\xi) - \beta^*(\xi))d\xi \right| \leq 2\beta_s^* \varepsilon \to 0 \text{ as } \varepsilon \to 0,$$

where we recall that $\beta_s^* = \lim_{r \nearrow y_s} \beta^*(r)$ (see (1.5)).
For $z \geq y_s$ we have

$$j_\varepsilon(z) = \int_0^{y_s - \varepsilon} \beta_\varepsilon^*(\xi)d\xi + \int_{y_s - \varepsilon}^z \beta_\varepsilon^*(\xi)d\xi = \int_0^{y_s - \varepsilon} \beta^*(\xi)d\xi$$

$$+ \beta^*(y_s - \varepsilon)[z - (y_s - \varepsilon)] + \frac{\beta_s^* - \beta^*(y_s - \varepsilon)}{2\varepsilon}[z - (y_s - \varepsilon)]^2.$$

Therefore, we have $\lim_{\varepsilon \to 0} j_\varepsilon(z) = j(y_s)$ for $z = y_s$ and

$$\lim_{\varepsilon \to 0} j_\varepsilon(z) = +\infty = j(z) \text{ for } z > y_s.$$

Now, we are going to show that

$$\int_{Q_\delta} j(y)dxdt \leq \liminf_{\varepsilon \to 0} \int_{Q_\delta} j_\varepsilon(y_\varepsilon)dxdt. \tag{1.88}$$

Let ε be small, e.g., $\varepsilon < \frac{y_s}{2}$. We can write

$$\int_{Q_\delta} j_\varepsilon(y_\varepsilon(t,x))dxdt \tag{1.89}$$

$$= \int_{Q_1^\varepsilon} j_\varepsilon(y_\varepsilon(t,x))dxdt + \int_{Q_2^\varepsilon} j_\varepsilon(y_\varepsilon(t,x))dxdt + \int_{Q_3^\varepsilon} j_\varepsilon(y_\varepsilon(t,x))dxdt,$$

where

$$Q_1^\varepsilon = \{(t,x) \in Q_\delta; \ y_\varepsilon(t,x) < y_s - \varepsilon\},$$
$$Q_2^\varepsilon = \{(t,x) \in Q_\delta; \ y_s - \varepsilon \leq y_\varepsilon(t,x) \leq y_s\},$$
$$Q_3^\varepsilon = \{(t,x) \in Q_\delta; \ y_s < y_\varepsilon(t,x)\}.$$

We compute each term apart. For $(t, x) \in Q_1^\varepsilon$ we have

$$j_\varepsilon(y_\varepsilon(t, x)) = \int_0^{y_\varepsilon(t,x)} \beta_\varepsilon^*(\xi)d\xi = \int_0^{y_\varepsilon(t,x)} \beta^*(\xi)d\xi = j(y_\varepsilon(t, x)).$$

For $(t, x) \in Q_2^\varepsilon$ we write

$$j_\varepsilon(y_\varepsilon(t, x)) = \int_0^{y_s-\varepsilon} \beta_\varepsilon^*(\xi)d\xi + \int_{y_s-\varepsilon}^{y_\varepsilon(t,x)} \beta_\varepsilon^*(\xi)d\xi$$

$$= j(y_s - \varepsilon) + \beta^*(y_s - \varepsilon)[y_\varepsilon(t, x) - (y_s - \varepsilon)]$$

$$+ \frac{\beta_s^* - \beta^*(y_s - \varepsilon)}{2\varepsilon}[y_\varepsilon(t, x) - (y_s - \varepsilon)]^2$$

$$\geq j(y_s - \varepsilon)$$

because the last two terms in the sum are positive on Q_2^ε (β^* is positive for a positive argument and so $\beta^*(y_s - \varepsilon) > 0$).

Next, if $(t, x) \in Q_3^\varepsilon$, taking into account that $\beta_\varepsilon^*(r) \geq \beta^*(r)$ for $r < y_s$ and $\beta_\varepsilon^*(y_s) = \beta_s^*$ we have

$$j_\varepsilon(y_\varepsilon(t, x)) = \int_0^{y_s} \beta_\varepsilon^*(\xi)d\xi + \int_{y_s}^{y_\varepsilon(t,x)} \beta_\varepsilon^*(\xi)d\xi$$

$$\geq \int_0^{y_s} \beta^*(\xi)d\xi + \beta^*(y_s - \varepsilon)(y_\varepsilon(t, x) - y_s)$$

$$+ \frac{\beta_s^* - \beta^*(y_s - \varepsilon)}{2\varepsilon}(y_\varepsilon(t, x) - y_s)^2$$

$$\geq j(y_s).$$

We resume (1.89), writing

$$\int_{Q_\delta} j_\varepsilon(y_\varepsilon(t, x))dxdt \geq \int_{Q_1^\varepsilon} j(y_\varepsilon(t, x))dxdt + \int_{Q_2^\varepsilon} j(y_s - \varepsilon)dxdt + \int_{Q_3^\varepsilon} j(y_s)dxdt$$

$$= \int_{Q_\delta} j(y(t, x))dxdt + \int_{Q_1^\varepsilon} (j(y_\varepsilon(t, x)) - j(y(t, x)))dxdt$$

$$+ \int_{Q_2^\varepsilon} (j(y_s - \varepsilon) - j(y(t, x)))dxdt + \int_{Q_3^\varepsilon} (j(y_s) - j(y(t, x)))dxdt \qquad (1.90)$$

and we treat again each term apart.

Since $y_\varepsilon \to y$ in $L^2(Q_\delta)$ it follows that on a subsequence $y_\varepsilon \to y$ a.e. on Q_δ, and in particular this is true on Q_1^ε and Q_2^ε. Moreover, $y \to j(y)$ is continuous if $y \leq y_s$ and so we have

$$j(y_\varepsilon(t, x)) - j(y(t, x)) \to 0 \text{ a.e. on } Q_1^\varepsilon, \text{ as } \varepsilon \to 0.$$

Then $j(y_\varepsilon(t,x)) \leq j(y_s - \varepsilon) \leq j(y_s)$ if $(t,x) \in Q_1^\varepsilon$ and so $|j(y_\varepsilon(t,x)) - j(y(t,x))| \leq 2j(y_s)$. In conclusion by the Lebesgue dominated convergence theorem we deduce that

$$\left| \int_{Q_1^\varepsilon} (j(y_\varepsilon(t,x)) - j(y(t,x)))dxdt \right|$$

$$= \int_{Q_\delta} \left| (j(y_\varepsilon(t,x)) - j(y(t,x)))\chi_{Q_1^\varepsilon}(t,x) \right| dxdt \to 0,$$

where $\chi_{Q_1^\varepsilon}$ is the characteristic function of the set Q_1^ε. For the second term in the sum (1.90) we write

$$\int_{Q_2^\varepsilon} (j(y_s - \varepsilon) - j(y(t,x)))dxdt$$

$$= \int_{Q_2^\varepsilon} (j(y_s - \varepsilon) - j(y_\varepsilon(t,x)))dxdt + \int_{Q_2^\varepsilon} (j(y_\varepsilon(t,x)) - j(y(t,x)))dxdt.$$

The last term on the right-hand side converges to 0 by a similar argument as before, using the Lebesgue dominated convergence theorem. For the first term we recall that $y \to j(y)$ is Lipschitz if $y \leq y_s$ and we have

$$\left| \int_{Q_2^\varepsilon} (j(y_s - \varepsilon) - j(y_\varepsilon(t,x)))dxdt \right| \leq \left| \int_{Q_\delta} (j(y_s - \varepsilon) - j(y_\varepsilon(t,x)))\chi_{Q_2^\varepsilon}(t,x)dxdt \right|$$

$$\leq \beta_s^* \int_{Q_\delta} |y_s - \varepsilon - y_\varepsilon(t,x)| \, dxdt$$

$$\leq \beta_s^* \text{meas}(Q_\delta)\varepsilon \to 0 \text{ as } \varepsilon \to 0,$$

where $\chi_{Q_2^\varepsilon}$ is the characteristic function of the set Q_2^ε.

For the third term in (1.90) we write

$$\int_{Q_3^\varepsilon} (j(y_s) - j(y(t,x)))dxdt = \int_{Q_\delta} (j(y_s) - j(y(t,x)))\chi_{Q_3^\varepsilon}(t,x)dxdt$$

where $\chi_{Q_3^\varepsilon}$ is the characteristic function of the set Q_3^ε.

We are going to show that

$$y(t,x) \leq y_s \text{ a.e. on } Q_\delta$$

which will imply that the integral on Q_3^ε is nonnegative.

Thus, on the basis of these results coming back to (1.90) we deduce

$$\liminf_{\varepsilon \to 0} \int_{Q_\delta} j_\varepsilon(y_\varepsilon(t,x)) dx dt \geq \int_{Q_\delta} j(y(t,x)) dx dt$$

$$+ \liminf_{\varepsilon \to 0} \left(\int_{Q_1^\varepsilon} (j(y_\varepsilon(t,x)) - j(y(t,x))) \, dx dt \right.$$

$$\left. + \int_{Q_2^\varepsilon} (j(y_s - \varepsilon) - j(y(t,x))) dx dt \right)$$

$$= \int_{Q_\delta} j(y(t,x)) dx dt$$

and so (1.88) is proved.

It remains to prove the assertion that $y(t,x) \leq y_s$ a.e. on Q_δ. We recall (1.66) which implies in particular

$$\int_0^t \|\beta_\varepsilon^* (y_\varepsilon(\tau))\|_{L^2(Q_\delta)}^2 \, d\tau \leq C$$

that can be still written

$$\int_0^t \|\beta_\varepsilon^* (y_\varepsilon(\tau))\|_{L^2(Q_3^\varepsilon)}^2 \, d\tau + \int_0^t \|\beta_\varepsilon^* (y_\varepsilon(\tau))\|_{L^2(Q_\delta \setminus Q_3^\varepsilon)}^2 \, d\tau \leq C.$$

The second term is positive and bounded, $\beta_\varepsilon^* (y_\varepsilon(\tau,x)) \leq \beta_s^*$ on $Q_\delta \setminus Q_3^\varepsilon = \{(t,x); y_\varepsilon(t,x) \leq y_s\}$, and replacing the expression of β_ε^* we obtain

$$\int_{Q_\delta} \left\{ \beta^*(y_s - \varepsilon) + \frac{\beta_s^* - \beta^*(y_s - \varepsilon)}{\varepsilon} [y_\varepsilon - (y_s - \varepsilon)] \right\}^2 \chi_{Q_3^\varepsilon}(t,x) dx dt \leq C.$$

Further we have

$$\int_{Q_\delta} \left(\frac{\beta_s^* - \beta^*(y_s - \varepsilon)}{\varepsilon} \right)^2 (y_\varepsilon - y_s)^2 \chi_{Q_3^\varepsilon}(t,x) dx dt \leq C$$

because $\beta^*(y_s - \varepsilon) > 0$. We recall that β^* is convex, which implies that

$$\frac{\beta_s^* - \beta^*(y_s - \varepsilon)}{\varepsilon} > \beta(y_s - \varepsilon)$$

and so we get

$$\int_{Q_\delta} (y_\varepsilon - y_s)^2 \chi_{Q_3^\varepsilon}(t,x) dx dt = \int_{Q_\delta} \{(y_\varepsilon - y_s)^+\}^2 dx dt \leq \frac{C}{\beta^2(y_s - \varepsilon)}$$

where $(y_\varepsilon - y_s)^+$ represents the positive part of $(y_\varepsilon - y_s)$. Now we pass to the limit (recalling that $y_\varepsilon \to y$ in $L^2(Q_\delta)$ by (1.84)) and take into account that β blows up at y_s, getting

$$\int_{Q_\delta} \{(y - y_s)^+\}^2 dx dt \leq 0$$

whence we deduce that $y(t, x) \leq y_s$ a.e. on Q_δ.

Now we resume the proof of the convergence of β_ε^* on Q_δ. Since

$$j_\varepsilon(r) \leq j_\varepsilon(z) + \beta_\varepsilon^*(r)(r - z), \text{ for any } r, z \in \mathbb{R},$$

we can write the inequality in particular for $z : (0, T) \times \Omega_\delta \to \mathbb{R}$, $z \in L^2(Q_\delta)$ and $r = y_\varepsilon$. We have

$$\int_{Q_\delta} j_\varepsilon(y_\varepsilon) dx dt \leq \int_{Q_\delta} j_\varepsilon(z) dx dt + \int_{Q_\delta} \beta_\varepsilon^*(y_\varepsilon)(y_\varepsilon - z) dx dt. \tag{1.91}$$

Assume $z \leq y_s$. Then $j_\varepsilon(z) \leq \beta_s^* y_s$ and using (1.87) we deduce by the Lebesgue dominated convergence theorem (see [13], pp. 3) that

$$\lim_{\varepsilon \to 0} \int_{Q_\delta} j_\varepsilon(z) dx dt = \int_{Q_\delta} j(z) dx dt.$$

Next, we remind that $\beta_\varepsilon^*(y_\varepsilon) \rightharpoonup \zeta$ in $L^2(0, T; V)$ and $y_\varepsilon \to y$ in $L^2(0, T; L^2(\Omega_\delta))$. By passing to limit as $\varepsilon \to 0$ in (1.91) and taking into account (1.88) we obtain that

$$\int_{Q_\delta} j(y) dx dt \leq \int_{Q_\delta} j(z) dx dt + \int_{Q_\delta} \zeta(y - z) dx dt, \ \forall z \in L^2(Q_\delta), \ z \leq y_s. \tag{1.92}$$

This implies that $\partial j = \zeta$. Here is the argument. Let us fix $(t_0, x_0) \in Q_\delta$, choose w arbitrary in $\mathbb{R}$, $w \leq y_s$, and define

$$z(t, x) := \begin{cases} y(t, x), & (t, x) \notin B_r(t_0, x_0) \\ w, & (t, x) \in B_r(t_0, x_0), \end{cases}$$

where $B_r(t_0, x_0)$ is the ball of centre (t_0, x_0) and radius $r > 0$. We denote $\overline{B}_r(t_0, x_0) = Q_\delta \backslash B_r(t_0, x_0)$. Then, (1.92) yields

$$\int_{B_r(t_0, x_0)} j(y) dx dt + \int_{\overline{B}_r(t_0, x_0)} j(y) dx dt$$

$$\leq \int_{B_r(t_0, x_0)} j(z) dx dt + \int_{\overline{B}_r(t_0, x_0)} j(z) dx dt$$

$$+ \int_{B_r(t_0, x_0)} \zeta(y - z) dx dt + \int_{\overline{B}_r(t_0, x_0)} \zeta(y - z) dx dt.$$

Taking into account the choice of $z(t,x)$ we have

$$\int_{B_r(t_0,x_0)} j(y)dxdt + \int_{\overline{B}_r(t_0,x_0)} j(y)dxdt$$

$$\leq \int_{B_r(t_0,x_0)} j(w)dxdt + \int_{\overline{B}_r(t_0,x_0)} j(y)dxdt$$

$$+ \int_{B_r(t_0,x_0)} \zeta(y-w)dxdt + \int_{\overline{B}_r(t_0,x_0)} \zeta(y-y)dxdt$$

from where it remains

$$\int_{B_r(t_0,x_0)} j(y)dxdt \leq \int_{B_r(t_0,x_0)} j(w)dxdt + \int_{B_r(t_0,x_0)} \zeta(y-w)dxdt.$$

We recall the following definition. Let l be a Lebesgue measurable function on a set S and let $z_0 \in S$. The point z_0 is called *a Lebesgue point* for l if

$$\lim_{r\to 0} \frac{1}{\text{meas}(B_r(z_0))} \int_{B_r(z_0)} l(x)dx = l(z_0).$$

The set of the points at which the previous relation holds is called the set of Lebesgue points. We also recall that the set of Lebesgue points for an integrable function l on a set S has the Lebesgue measure equal to that of S, namely almost all points in S are Lebesgue for l.

Thus, let us assume now that (t_0, x_0) considered before is a Lebesgue point for j. Dividing the inequality by $\text{meas}(B_r(x_0, t_0))$ and letting $r \to 0$ we get

$$j(y(t_0, x_0)) \leq j(w) + \zeta(t_0, x_0)(y(t_0, x_0) - w), \quad \forall w \in \mathbb{R}, \ w \leq y_s.$$

By the definition of j we get $\zeta(t,x) \in \beta^*(y(t,x))$ a.e. $(t,x) \in Q_\delta$. Then, since δ is arbitrary and $Q_u = \bigcup_{\delta>0} Q_\delta$, we infer that

$$\zeta(t,x) \in \beta^*(y(t,x)) \text{ a.e. on } Q_u,$$

and we deduce that

$$y(t,x) \leq y_s \text{ a.e. on } Q_u.$$

Finally, since $\left(K\left(\frac{\theta_\varepsilon}{u_\varepsilon}\right)\right)_\varepsilon$ is bounded in $L^2(Q)$ we have

$$K\left(\frac{\theta_\varepsilon}{u_\varepsilon}\right) \rightharpoonup \kappa \text{ in } L^2(Q), \text{ as } \varepsilon \to 0$$

and we assert that $\kappa = K(y)$. Indeed,

$$K\left(\frac{\theta_\varepsilon}{u_\varepsilon}\right) \rightharpoonup \kappa \text{ in } L^2(Q_u), \text{ as } \varepsilon \to 0,$$

too. On the other hand, K being Lipschitz it follows by (1.84) that $\left(K\left(\frac{\theta_\varepsilon}{u_\varepsilon}\right)\right)_{\varepsilon>0}$ is strongly convergent on each subset Q_δ,

$$K\left(\frac{\theta_\varepsilon}{u_\varepsilon}\right) \to K\left(y\right) \text{ in } L^2(Q_\delta), \text{ as } \varepsilon \to 0.$$

By the uniqueness of the limit the restriction of the weak limit function κ to Q_δ must coincide with $K(y)$ and this also implies that

$$\kappa(t,x) = K(y(t,x)) \text{ a.e. on } Q_u.$$

On the subset Q_0 the function $a(x)K\left(\frac{\theta_\varepsilon}{u_\varepsilon}\right) = 0$, so by the definition of K_0 we get

$$K_0\left(x,\frac{\theta_\varepsilon}{u_\varepsilon}\right) \rightharpoonup K_0(x,y) \text{ in } L^2(Q), \text{ as } \varepsilon \to 0.$$

Finally, we derive a relation which will serve a little later. Assume first that $f \in W^{1,2}([0,T]; L^2(\Omega))$ and $\theta_0 \in V$.

We recall that $\zeta(t) \in V$ a.e. $t \in (0,T)$. Since this regularity is not sufficient to define its normal derivative to a surface $\Gamma_c \subset \Omega$, we define a generalized normal derivative of it $\frac{\partial \zeta(t)}{\partial \nu}$, as an element of a distribution space on Γ_c. As a matter of fact $\frac{\partial \zeta(t)}{\partial \nu} \in H^{-1/2}(\Gamma_c)$ which is the dual of $H^{1/2}(\Gamma_c)$ (see the definitions of these spaces in [78]).

Assume that Γ_c is a smooth surface surrounding the domain $\Omega_c \subset \Omega$, i.e., $\Gamma_c = \partial \Omega_c$. If $\eta \in H^1(\Omega_c)$ and $\Delta \eta \in (H^1(\Omega_c))'$ then we define $\frac{\partial \eta}{\partial \nu} \in H^{-1/2}(\Gamma_c)$ by

$$\left\langle \frac{\partial \eta}{\partial \nu}, tr(\psi) \right\rangle_{H^{-1/2}(\Gamma_c), H^{1/2}(\Gamma_c)}$$

$$= \langle \Delta \eta, \psi \rangle_{(H^1(\Omega_c))', H^1(\Omega_c)} + \int_{\Omega_c} \nabla \eta \cdot \nabla \psi dx, \quad \forall \psi \in H^1(\Omega_c). \qquad (1.93)$$

In particular, for $\eta = \zeta(t)$, $\Omega_c = \Omega_0$ with the boundary $\Gamma_0 = \partial \Omega_0$ we define the outward normal derivative $\frac{\partial^+}{\partial \nu}\zeta(t)$ a.e. $t \in (0,T)$, by

$$\left\langle \frac{\partial^+ \zeta(t)}{\partial \nu}, tr(\psi) \right\rangle_{H^{-1/2}(\Gamma_0), H^{1/2}(\Gamma_0)}$$

$$= \langle \Delta \zeta(t), \psi \rangle_{(H^1(\Omega_0))', H^1(\Omega_0)}$$

$$+ \int_{\Omega_0} \nabla \zeta(t) \cdot \nabla \psi dx, \quad \forall \psi \in H^1(\Omega_0), \text{ a.e. } t \in (0,T), \qquad (1.94)$$

where $tr(\psi)$ is the trace of $\psi \in H^1(\Omega_0)$ on Γ_0.

In a similar way, considering $\Omega_u = \Omega \backslash \Omega_0$ which has the common boundary $\Gamma_0 = \partial \Omega_0$ with Ω_0, we define $\frac{\partial^-}{\partial \nu} \zeta(t)$ on Γ_0 by the relation

$$\left\langle \frac{\partial^- \zeta(t)}{\partial \nu}, tr(\psi) \right\rangle_{H^{-1/2}(\Gamma_0), H^{1/2}(\Gamma_0)}$$

$$= \langle \Delta \zeta(t), \psi \rangle_{(H^1(\Omega_u))', H^1(\Omega_u)} + \int_{\Omega_u} \nabla \zeta(t) \cdot \nabla \psi dx, \ \forall \psi \in V, \ \text{a.e. } t \in (0, T),$$

$$(1.95)$$

where $tr(\psi)$ is the trace of $\psi \in V$ on Γ_0.

Thus we can obtain the continuity of the generalized normal derivative across the surface Γ_c, in particular across Γ_0. Indeed by (1.65) we have

$$\int_{\Gamma_0} (K_0^+(x, y_\varepsilon(t)) - \nabla \zeta_\varepsilon^+(t)) \cdot \nu^+ \psi d\sigma$$

$$= \int_{\Gamma_0} (K_0^-(x, y_\varepsilon(t)) - \nabla \zeta_\varepsilon^-(t)) \cdot \nu^- \psi d\sigma, \ \forall \psi \in V, \ \text{a.e. } t \in (0, T),$$

where $\zeta_\varepsilon = \beta_\varepsilon^*(y_\varepsilon)$ and the superscripts $+$ and $-$ denote the restrictions of the functions on Ω_0 and Ω_u, respectively. Also, ν^+ and ν^- are the outer normal derivatives to Γ_0 from Ω_0 and Ω_u, respectively. Since $\nabla \zeta_\varepsilon(t)$ is bounded in $L^2(\Omega)$ independently on ε, a.e. t (see (1.52)) we can pass to the limit and get

$$\left\langle (K_0^+(\cdot, y(t)) - \nabla \zeta^+(t)) \cdot \nu^+, \psi \right\rangle_{H^{-1/2}(\Gamma_0), H^{1/2}(\Gamma_0)}$$

$$= \left\langle K_0^-(\cdot, y(t)) - \nabla \zeta^-(t) \cdot \nu^-, \psi \right\rangle_{H^{-1/2}(\Gamma_0), H^{1/2}(\Gamma_0)}, \ \forall \psi \in V, \ \text{a.e. } t \in (0, T)$$

$$(1.96)$$

where the normal derivatives $\frac{\partial^+ \zeta(t)}{\partial \nu} = \nabla \zeta^+(t) \cdot \nu^+$ and $\frac{\partial^- \zeta(t)}{\partial \nu} = \nabla \zeta^-(t) \cdot \nu^-$ are considered in the generalized sense (1.94) and (1.95). For simplicity, here we denoted $tr(\psi)$ still by ψ.

Now we can pass to limit as $\varepsilon \to 0$ in (1.43) and obtain

$$\int_0^T \left\langle \frac{d(uy)}{dt}(t), \phi(t) \right\rangle_{V', V} dt + \int_Q (\nabla \zeta - K_0(x, y)) \cdot \nabla \phi dx dt$$

$$= \int_0^T \int_\Omega f \phi dx dt, \quad \text{for any } \phi \in L^2(0, T; V), \tag{1.97}$$

where ζ is given by (1.69), $\zeta = \lim_{\varepsilon \to 0} \beta_\varepsilon^*(y_\varepsilon)$.

In particular if $\phi \in C_0^\infty(Q_u)$ we get

$$\int_0^T \left\langle \frac{d(uy)}{dt}(t), \phi(t) \right\rangle_{V',V} dt + \int_{Q_u} (\nabla\zeta - K_0(x,y)) \cdot \nabla\phi \, dx dt$$

$$= \int_0^T \int_{\Omega_u} f\phi \, dx dt, \tag{1.98}$$

where $\zeta \in \beta^*(y)$ a.e. on Q_u. We have taken into account that

$$\frac{d(uy)}{dt} = \begin{cases} \frac{\partial(uy)}{\partial t}, & \text{if } uy > 0 \\ 0, & \text{if } uy = 0 \end{cases} \tag{1.99}$$

where $\frac{\partial(uy)}{\partial t}$ is the derivative in the sense of distributions.

If we take $\phi \in C_0^\infty(Q_0)$ we obtain

$$\int_{Q_u} (\nabla\zeta - K_0(x,y)) \cdot \nabla\phi \, dx dt = \int_0^T \int_{\Omega_0} f\phi \, dx dt, \tag{1.100}$$

where ζ is given by (1.69).

1.1.6   Construction of the Solution

Now we consider the following equations in the sense of distributions

$$\frac{\partial(uy)}{\partial t} - \Delta\zeta + \nabla \cdot K_0(x,y) \ni f \quad \text{in } Q,$$

$$\zeta = 0 \quad \text{on } \Sigma, \tag{1.101}$$

obtained from (1.97) for $\phi \in C_0^\infty(Q)$, where ζ is given by (1.69),

$$\frac{\partial(uy)}{\partial t} - \Delta\zeta + \nabla \cdot K_0(x,y) \ni f \quad \text{in } Q_u = (0,T) \times \Omega_u,$$

$$\zeta = 0 \quad \text{on } \Sigma, \tag{1.102}$$

with $\zeta(t,x) \in \beta^*(y(t,x))$ a.e. $(t,x) \in Q_u$ and

$$-\Delta\zeta \ni f \quad \text{in } Q_0 = (0,T) \times \Omega_0 \tag{1.103}$$

with ζ given again by (1.69).

The common boundary $\partial\Omega_0$ of the domains Ω_u and Ω_0 is regular. Since $\zeta \in L^2(0,T;V)$ we deduce that for a.e. $t \in (0,T)$ the trace of the function

$\zeta(t)$ on any line $\mathcal{L}_0 \subset \Omega$ crossing the boundary $\partial\Omega_0$ belongs to V, so that it is continuous across $\mathcal{L}_0$. Thus if we take $x_0 \in \partial\Omega_0$ then

$$\zeta^-(t) := \lim_{\substack{x \to x_0 \\ x \in \mathcal{L}_0 \cap \Omega_u}} \zeta(t) = \lim_{\substack{x \to x_0 \\ x \in \mathcal{L}_0 \cap \Omega_0}} \zeta(t) = \zeta^+(t) \text{ a.e. } t \in (0,T).$$

We take into account that $\zeta^-(t) \in \beta^*(y(t))$ a.e. $t \in (0,T)$, hence ζ turns out to be the solution to the elliptic problem

$$-\Delta\zeta(t) = f(t) \qquad\qquad \text{in } \Omega_0, \text{ a.e. } t \in (0,T) \qquad (1.104)$$

$$\zeta(t) = \zeta^-(t) \in \beta^*(y(t)) \qquad \text{on } \partial\Omega_0, \text{ a.e. } t \in (0,T),$$

where y is the solution to (1.102) in Q_u.

Now we can construct the function

$$y^*(t,x) := \begin{cases} y(t,x), & \text{if } (t,x) \in Q_u \\ (\beta^*)^{-1}(\zeta(t,x)), & \text{if } (t,x) \in Q_0 \end{cases} \qquad (1.105)$$

and show that it is the solution to (1.27). Since $\zeta \in L^2(0,T;V)$ it follows that $y^* \in L^2(0,T;D(A))$, whence $y^* \leq y_s$ a.e. on Q. This function belongs also to the spaces specified in (1.23) (for the derivative we take into account (1.99)).

We have to check that y^* satisfies (1.26). If we plug y^* given by (1.105) in (1.26) and we take into account (1.99), (1.96) we obtain

$$\int_0^T \left\langle \frac{d(uy^*)}{dt}(t), \phi(t) \right\rangle_{V',V} dt + \int_Q (\nabla\zeta - K_0(x,y^*)) \cdot \nabla\phi \, dx dt$$

$$= \int_0^T \left\langle \frac{d(uy^*)}{dt}(t), \phi(t) \right\rangle_{V',V} dt + \int_{Q_u} (\nabla\zeta - K_0(x,y)) \cdot \nabla\phi \, dx dt$$

$$+ \int_{Q_0} (\nabla\zeta - K_0(x,y^*)) \cdot \nabla\phi \, dx dt$$

$$= \int_0^T \int_{\Omega_u} f\phi \, dx dt + \int_0^T \int_{\Omega_0} f\phi \, dx dt = \int_0^T \int_\Omega f\phi \, dx dt,$$

for any $\phi \in L^2(0,T;V)$, $\zeta \in \beta^*(y^*)$ a.e. on Q. Here we used (1.98) and (1.100).

Now, let $f \in L^2(0,T;V')$ and $\theta_0 \in L^2(\Omega)$. The previous relation remains true, by density, but we do not provide all arguments because they are similar with those given up to now. So, we obtain (1.26) as claimed and this ends the existence proof. $\qquad\square$

Now we are going to specify a physical interpretation of the solution, stating that the previous proof also implies

Corollary 1.7. *The solution y^* to problem (1.1) given by Theorem 1.6 is the solution to the transmission problem*

$$\frac{\partial(u(x)y^*)}{\partial t} - \Delta\beta^*(y^*) + \nabla \cdot K_0(x, y^*) \ni f \qquad in \ Q_u,$$

$$-\Delta\beta^*(y^*) \ni f \qquad in \ Q_0,$$

$$\zeta^+ = \zeta^- \qquad on \ \Sigma_0 = (0, T) \times \partial\Omega_0,$$

$$(K_0^+(x, y^*) - \nabla\zeta^+) \cdot \nu^+ = (K_0^-(x, y^*) - \nabla\zeta^-) \cdot \nu^+ \qquad on \ \Sigma_0,$$

$$y^*(t, x) = 0 \qquad on \ \Sigma := (0, T) \times \Gamma,$$

$$(u(x)y^*(t, x))|_{t=0} = \theta_0(x) \qquad in \ \Omega. \tag{1.106}$$

Proof. Let $f \in W^{1,2}([0, T]; L^2(\Omega))$. Let us write that y^* is a solution to (1.1)

$$\int_0^T \left\langle \frac{d(uy^*)}{dt}(t), \phi(t) \right\rangle_{V', V} dt + \int_{Q_u} (\nabla\zeta - K_0(x, y)) \cdot \nabla\phi \, dx dt$$

$$+ \int_{Q_0} (\nabla\zeta - K_0(x, y^*)) \cdot \nabla\phi \, dx dt$$

$$= \int_0^T \int_\Omega f\phi \, dx dt,$$

whence, expressing the integrals on Q_u and Q_0 in another way, we get

$$\int_0^T \left\langle \frac{d(uy^*)}{dt}(t) - \Delta\zeta(t) + \nabla \cdot a(x)K_0(x, y^*(t)) - f(t), \phi(t) \right\rangle_{(H^1(\Omega_u))', H^1(\Omega_u)} dt$$

$$- \int_0^T \left\langle (K_0^-(\cdot, y^*(t)) - \nabla\zeta^-(t)) \cdot \nu^-, \phi(t) \right\rangle_{H^{-1/2}(\partial\Omega_0), H^{1/2}(\partial\Omega_0)} dt$$

$$+ \int_0^T \left\langle -\Delta\zeta(t) - f(t), \phi(t) \right\rangle_{(H^1(\Omega_0))', H^1(\Omega_0)} dt$$

$$- \int_0^T \left\langle K_0^+(\cdot, y^*(t)) - \nabla\zeta^+(t) \cdot \nu^+, \phi(t) \right\rangle_{H^{-1/2}(\partial\Omega_0), H^{1/2}(\partial\Omega_0)} dt$$

$$= 0,$$

for any $\phi \in C_0^\infty(Q)$. Using (1.102) and (1.103) we get

$$(K_0^-(\cdot, y^*(t)) - \nabla\zeta^-(t))) \cdot \nu^- + (K_0^+(\cdot, y^*(t)) - \nabla\zeta^+(t)) \cdot \nu^+ = 0 \text{ a.e. } t, \text{ on } \partial\Omega_0$$

where $\nu^- = -\nu^+$. The result remains true for $f \in L^2(0, T; V')$, by density. $\square$

This means that the flux is conserved across the boundary Σ_0, which from the physical point of view is natural. As a matter of fact (1.106) is an equivalent form of (1.27).

Finally, we mention that the presence of the advection term in nonlinear degenerate diffusion problems, as well as in periodic problems as we shall see, may induce difficulties in proving the solution uniqueness, especially when using energetic relations. This is not a singular situation, because as it is well known there are many nonlinear problems in which uniqueness has remained an open problem (e.g. Navier–Stokes equation in 3D, nonlinear wave equation). In general uniqueness follows under restrictive assumptions and in diffusion with transport problems one can observe that it is ensured when the diffusion dominates the advection. In media with low porosity it can also be shown that a small enough velocity of the fluid is a condition guaranteeing the flow uniqueness. So, we give next a uniqueness result, establishing in fact a sufficient condition in (1.107) below. Its interpretation is that the advection vector in absolute value is of the same order of magnitude as the square root of the porosity. For the case when (1.107) is not obeyed one can accept that the approximating solution (which is unique) is an appropriate candidate for the solution to the physical model (1.1).

Proposition 1.8. *Under the hypotheses of Theorem 1.6 assume in addition that there exists $k_u > 0$ such that*

$$|a(x)| \leq k_u \sqrt{u(x)} \text{ for any } x \in \Omega. \tag{1.107}$$

Then the solution to (1.1) is unique a.e. on Q.

Proof. Assume that we have two solutions (y^*, ζ) and $(\overline{y}^*, \overline{\zeta})$ to (1.27) corresponding to the same data f and θ_0. We subtract (1.27) written for y^* and $\overline{y}^*$, multiply the difference scalarly in V' by $u(y^* - \overline{y}^*)(t)$, and integrate over $(0, t)$. We get

$$\int_0^t \left(\frac{d(u(y^* - \overline{y}^*))}{d\tau}(\tau), u(y^* - \overline{y}^*)(\tau) \right)_{V'} d\tau + \int_0^t \int_\Omega \nabla(\zeta - \overline{\zeta}) \cdot \nabla\psi dx d\tau$$

$$= \int_0^t \int_\Omega (K(y^*) - K(\overline{y}^*))a(x) \cdot \nabla\psi dx d\tau, \tag{1.108}$$

where $A_0\psi = u(y^* - \overline{y}^*)$. Next we have

$$\frac{1}{2} \|u(y^* - \overline{y}^*)(t)\|_{V'}^2 + \int_0^t \int_\Omega (\zeta - \overline{\zeta})u(y^* - \overline{y}^*) dx d\tau$$

$$\leq NM_K k_u \int_0^t \int_\Omega |\sqrt{u}(y^* - \overline{y}^*)| |\nabla\psi| dx d\tau$$

$$\leq NM_K k_u \int_0^t \|\sqrt{u}(y^* - \overline{y}^*)(\tau)\| \|u(y^* - \overline{y}^*)(\tau)\|_{V'} d\tau$$

whence, recalling (1.7) we obtain

$$\frac{1}{2} \|u(y^* - \overline{y}^*)(t)\|_{V'}^2 + \rho \int_0^t \int_\Omega u(y^* - \overline{y}^*)^2 dx d\tau$$

$$\leq \frac{\rho}{2} \int_0^t \int_\Omega u(y^* - \overline{y}^*)^2 dx d\tau + \frac{1}{2\rho}(N M_K k_u)^2 \int_0^t \|u(y^* - \overline{y}^*)(\tau)\|_{V'}^2 d\tau.$$

Therefore, by Gronwall lemma (see [29]), $\|u(y^* - \overline{y}^*)(t)\|_{V'}^2 \leq 0$ and we deduce that $uy^*(t) = u\overline{y}^*(t)$ for any $t \in [0, T]$. It follows that the solution is unique a.e. on the set Q_u where $u(x) > 0$. Therefore, using (1.104) which is satisfied by $\zeta(t) \in \beta^*(y^*(t))$ and $\overline{\zeta}(t) \in \beta^*(y^*(t))$ we write the problem satisfied by their difference

$$\Delta(\zeta - \overline{\zeta})(t) = 0 \text{ in } \Omega_0, \text{ a.e. } t \in (0, T),$$

$$(\zeta - \overline{\zeta})(t) = 0 \text{ on } \partial\Omega_0, \text{ a.e. } t \in (0, T).$$

This implies that $\zeta(t) = \overline{\zeta}(t)$ a.e. t and since $(\beta^*)^{-1}$ is single valued we get that $y^*(t) = \overline{y}^*(t)$ a.e. on Ω_0. Then the solution uniqueness follows a.e. on Q. $\square$

Finally we would like to make a short comment about the continuity of the solution with respect to the nonlinear functions, without entering into details. We recall that such a property has been studied in [25] in the case of Richards' equation.

First we focus on the approximating problem (1.40). Let $(K_j)_j$ be such that $K_j(r) \to K(r)$ as $j \to \infty$, and $(\beta_j^*)_j$ be a family of graphs such that $(\beta_j^*)_j$ converges to β^* in the sense of the resolvent, that is

$$(1 + \lambda\beta_j^*)^{-1} z \to (1 + \lambda\beta^*)^{-1} z, \text{ as } j \to \infty, \ \forall \lambda > 0, \ \forall z \in \mathbb{R}.$$

Then

$$(I + \lambda B_\varepsilon^j)^{-1} g \to (I + \lambda B_\varepsilon)^{-1} g \text{ as } j \to \infty, \text{ for } g \in V',$$

where B_ε^j are the quasi m-accretive operators in (1.40) corresponding to $(\beta_j^*)_\varepsilon$. Then by Trotter–Kato theorem for nonlinear semigroups (see [14], pp. 168) it follows that the corresponding sequence of solutions $(\theta_j)_\varepsilon$ is convergent to θ_ε as $j \to \infty$ in $C([0, T]; V')$. This continuity result can be further used to get the continuity for the solution to the limit equation when $\varepsilon \to 0$.

1.1.7 Numerical Results

We end this chapter with numerical simulations for the solution to (1.1). We imagine some scenarios for a real-world model of water infiltration into a nonhomogeneous porous medium (soil) in which a solid intrusion with zero porosity (a rock) is present. Assuming that the model (1.1) is already written in a dimensionless form, let us consider the expressions

$$\beta(r) = \frac{c(c-1)}{(c-r)^2}, \quad K(r) = \frac{(c-1)r^2}{c-r} \quad \text{for } r \in [0,1), \quad c > 1, \qquad (1.109)$$

given by the parametric model of Broadbridge and White (see [33]). These functions characterize the water infiltration into a soil whose properties are strongly nonlinear when c is in a neighborhood of 1 and weakly nonlinear for larger values of c (e.g., $c \geq 1.2$).

We see that here $\lim_{r \to y_s=1} \beta(r)$ is finite. This may be obtained by a jump of the function C (defined in Introduction) at $r = r_s = 0$ from a positive value at the left to 0 at the right (see case (a) in Introduction), such that the function β^* is multivalued at $r = 1$. All the results proved in this section apply to this case as well.

The computations are done in the 2D case in the domain

$$\Omega = \{(x_1, x_2); x_1 \in (0,5), \ x_2 \in (0,5)\},$$

with Ω_0 the circle with center in $(2,3)$ and radius $\delta = 0.1$,

$$\Omega_0 = \{(x_1, x_2); (x_1 - 2)^2 + (x_2 - 3)^2 \leq 0.1^2\}$$

and the function u (expressing the porosity of the soil) is chosen of the form

$$u(x_1, x_2) := \begin{cases} 0, & \text{in } \Omega_0 \\ \frac{(x_1-2)^2+(x_2-3)^2-0.1^2}{100}, & \text{in } \Omega_u. \end{cases} \qquad (1.110)$$

In the computations we take $u_\varepsilon(x_1, x_2) = u(x_1, x_2) + 10^{-9}$.

The functions β^* and K_0 with the properties considered in this section are

$$\beta^*(r) = \begin{cases} \frac{(c-1)r}{c-r}, & r \in [0,1) \\ [1, \infty), & r = 1, \end{cases} \qquad K_0(x, r) = \begin{cases} a(x) \begin{cases} \frac{(c-1)r^2}{c-r}, & r \in [0,1) \\ 1, & r \geq 1 \end{cases} & \text{in } \Omega_0 \\ 0, & \text{in } \Omega_u. \end{cases}$$

The other data are: $\theta_0(x_1, x_2) = 0$, $a(x_1, x_2) = (1,1)$, meaning that the initial soil is dry and the advection is along both directions, and

$$f(t, x_1, x_2) = \begin{cases} t^2, & \text{in } \Omega_u \\ 0, & \text{in } \Omega_0. \end{cases}$$

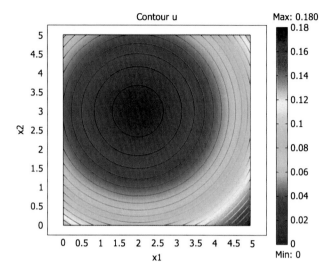

Fig. 1.2 Contour plot of the function u given by (1.110)

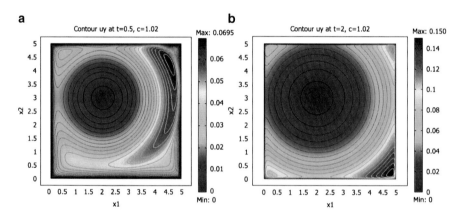

Fig. 1.3 Solution $\theta = uy$ in the parabolic–elliptic degenerate case for u given by (1.110) and $c = 1.02$

The algorithm is adapted from [39] for this degenerate case and the computations are done by using the software package Comsol Multiphysics (see [40]).

In Fig. 1.2 it is represented the contour plot of the function $x_3 = u(x_1, x_2)$, i.e., the projection of this surface on the plane x_1Ox_2.

We are interested in some comparisons. In Fig. 1.3a, b we see the evolution of $\theta = uy$ (representing the volumetric water content or soil moisture) computed for $c = 1.02$ (a strongly nonlinear soil) at two moments of time

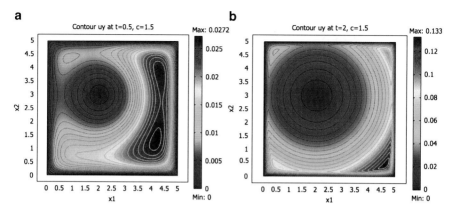

Fig. 1.4 Solution $\theta = uy$ in the parabolic–elliptic degenerate case for u given by (1.110) and $c = 1.5$

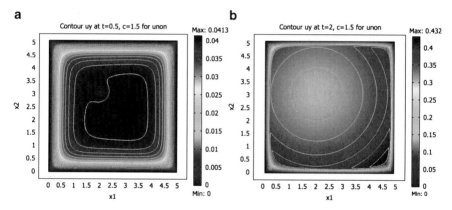

Fig. 1.5 Solution θ in the parabolic–elliptic nondegenerate case for u given by (1.111) and $c = 1.5$

$t = 0.5$ (Fig. 1.3a) and $t = 2$ (Fig. 1.3b), while in Fig. 1.4a, b we see the evolution of θ computed for $c = 1.5$ (a weakly nonlinear soil).

Then we compare the graphics in Fig. 1.4a, b with those drawn in Fig. 1.5a, b corresponding to the nondegenerate case with u positive given by the relation

$$unon(x_1, x_2) = u(x_1, x_2) + 0.3 \tag{1.111}$$

and $c = 1.5$. This describes a porous medium with a higher porosity which does not vanish, in which we see that the volumetric water content θ can reach higher values than in porosity vanishing case.

1.2 Well-Posedness for the Cauchy Problem with Very Fast Diffusion

Let us consider the problem

$$\frac{\partial(u(x)y)}{\partial t} - \Delta\beta^*(y) + \nabla \cdot K_0(x,y) = f(t,x) \quad \text{in } Q,$$

$$y(t,x) = 0 \quad \text{on } \Sigma, \qquad (1.112)$$

$$(u(x)y(t,x))|_{t=0} = \theta_0(x) \quad \text{in } \Omega,$$

in which β^* is a single valued function, β and β^* blow-up at $r = y_s$,

$$\lim_{r \nearrow y_s} \beta(r) = +\infty, \quad \lim_{r \nearrow y_s} \beta^*(r) = \lim_{r \nearrow y_s} \int_0^r \beta(s)ds = +\infty \qquad (1.113)$$

(see case (b) in Introduction) and

$$\beta(r) = \rho > 0, \text{ for any } r \le 0.$$

The functions u, a_i and K are assumed to be as in the fast diffusion case, i.e., obeying (1.10)–(1.14).

In this case we introduce the function $j : \mathbb{R} \to (-\infty, +\infty]$ by

$$j(r) := \begin{cases} \int_0^r \beta^*(\xi)d\xi, & r < y_s, \\ +\infty, & r \ge y_s, \end{cases}$$

and specify that j is proper, convex, l.s.c. and

$$\partial j(r) = \begin{cases} \beta^*(r), & r < y_s, \\ +\infty, & r \ge y_s, \end{cases}$$

(see the proof in [84], pp. 74).

Let us assume that

$$f \in L^2(0, T; V'), \qquad (1.114)$$

$$\theta_0 \in L^2(\Omega), \ \theta_0 = 0 \text{ a.e. on } \Omega_0, \qquad (1.115)$$

$$\theta_0 \ge 0 \text{ a.e. on } \Omega_u, \ \frac{\theta_0}{u} \in L^2(\Omega_u), \ j\left(\frac{\theta_0}{u}\right) \in L^1(\Omega).$$

Definition 1.9. Let (1.114) and (1.115) hold. We call a *weak solution* to (1.112) a function y such that

$$y \in L^2(0,T;V), \ \beta^*(y) \in L^2(0,T;V),$$
$$uy \in C([0,T];L^2(\Omega)) \cap W^{1,2}([0,T];V'),$$

which satisfies

$$\left\langle \frac{d(uy)}{dt}(t), \psi \right\rangle_{V',V} + \int_\Omega (\nabla\beta^*(y)(t) - K_0(x,y(t))) \cdot \nabla\psi \, dx$$
$$= \langle f(t), \psi \rangle_{V',V}, \ \text{a.e. } t \in (0,T), \ \text{for any } \psi \in V,$$

the initial condition $(uy(t))|_{t=0} = \theta_0$ and the boundedness condition

$$y(t,x) < y_s \text{ a.e. } (t,x) \in Q.$$

In the same way as in the previous section we can write the abstract Cauchy problem

$$\frac{d(uy)}{dt}(t) + Ay(t) = f(t), \ \text{a.e. } t \in (0,T), \tag{1.116}$$
$$(uy(t))|_{t=0} = \theta_0,$$

where

$$D(A) := \left\{ y \in L^2(\Omega); \ \beta^*(y) \in V \right\}$$

and $V = H_0^1(\Omega)$, with the dual $V' = H^{-1}(\Omega)$.

Then we pass to (1.30) by denoting $\theta(t,x) = u(x)y(t,x)$.

Next we shall prove that (1.116) has a weak solution.

Theorem 1.10. *Let us assume (1.114) and (1.115). Then, the Cauchy problem (1.116) has at least a weak solution y^*. In addition, if (1.107) holds, then the solution is unique.*

Proof. The proof is led as in the case of fast diffusion, with some modifications imposed by the blowing-up of β^*. First, we introduce the approximating functions β_ε and β_ε^* by

$$\beta_\varepsilon(r) := \begin{cases} \beta(r), & r < y_s - \varepsilon \\ \beta(y_s - \varepsilon), & r \geq y_s - \varepsilon, \end{cases} \tag{1.117}$$

$$\beta_\varepsilon^*(r) := \begin{cases} \beta^*(r), & r < y_s - \varepsilon \\ \beta^*(y_s - \varepsilon) + \beta(y_s - \varepsilon)[r - (y_s - \varepsilon)], & r \geq y_s - \varepsilon \end{cases} \tag{1.118}$$

and the approximating problem (1.40). It has a unique strong solution satisfying estimate (1.52), by using the same arguments as in Proposition 1.5. Then, if $j\left(\frac{\theta_0}{u}\right) \in L^1(\Omega)$ one can see that the upper bound of this estimate does not depend on ε and the proof can be continued as in Theorem 1.6.

The delicate point is to show the convergence of $\beta_\varepsilon^*(y_\varepsilon)$ to $\beta^*(y)$ in $L^2(0,T;L^2(\Omega_u))$. This is implied by the convergencies (1.84), (1.85)

$$y_\varepsilon \to y \text{ in } L^2(0,T;L^2(\Omega_\delta)) \text{ as } \varepsilon \to 0,$$

$$y_\varepsilon \rightharpoonup y \text{ in } L^2(0,T;L^2(\Omega_u)) \text{ as } \varepsilon \to 0$$

and (1.69)

$$\beta_\varepsilon^*\left(\frac{\theta_\varepsilon}{u_\varepsilon}\right) \rightharpoonup \zeta \text{ in } L^2(0,T;V), \text{ as } \varepsilon \to 0. \tag{1.119}$$

We claim that $\zeta = \beta^*(y)$ a.e. on Q_u. For this we set

$$Q_{\delta s} := \{(t,x) \in Q_\delta; \ y(t,x) = y_s\}, \ Q_{\delta n} := \{(t,x) \in Q_\delta; \ y(t,x) < y_s\}.$$

Then, if $(t,x) \in Q_{\delta n}$ we have $\beta_\varepsilon(r) = \beta(r)$ (for ε small enough) and we can write

$$\beta_\varepsilon^*(y_\varepsilon(t,x)) = \int_0^{y_\varepsilon(t,x)} \beta_\varepsilon(r)dr = \int_0^{y_\varepsilon(t,x)} \beta(r)dr$$

$$\to \int_0^{y(t,x)} \beta(r)dr = \beta^*(y(t,x)) \text{ a.e. on } Q_{\delta n}, \text{ as } \varepsilon \to 0.$$

If $(t,x) \in Q_{\delta s}$, then two situations may arise:

(p$_1$) there is a sequence $\varepsilon_k \to 0$ such that $y_{\varepsilon_k}(t,x) \geq y_s - \varepsilon_k$.
(p$_2$) for all $\varepsilon < \varepsilon_0$ we have $y_\varepsilon(t,x) < y_s - \varepsilon$.

In the case (p$_2$) the previous argument for $(t,x) \in Q_{\delta n}$ applies and $\beta_\varepsilon^*(y_\varepsilon) \to \beta^*(y)$ a.e. for $(t,x) \in Q_{\delta s}$.
In the case (p$_1$) we have

$$\beta_{\varepsilon_k}^*(y_{\varepsilon_k}(t,x)) = \int_0^{y_{\varepsilon_k}(t,x)-\varepsilon_k} \beta(r)dr + \int_{y_{\varepsilon_k}(t,x)-\varepsilon_k}^{y_{\varepsilon_k}(t,x)} \beta\left(y_s - \varepsilon_k\right)dr$$

$$= \int_0^{y_{\varepsilon_k}(t,x)-\varepsilon_k} \beta(r)dr + \varepsilon_k\beta(y_s - \varepsilon_k) \to +\infty = \beta^*(y_s),$$

as $\varepsilon_k \to 0$,

because $\int_0^{y_s} \beta(r)dr = +\infty$, pursuant to (1.113). Hence, selecting a subsequence (denoted still by the subscript ε), we have that

$$\beta_\varepsilon^*(y_\varepsilon) \to \beta^*(y) \text{ a.e. on } Q_\delta \text{ as } \varepsilon \to 0.$$

But $(\beta_\varepsilon^*(y_\varepsilon))_{\varepsilon>0}$ is bounded in $L^2(Q_\delta)$ by (1.66) and since it converges a.e. on Q_δ, it follows that $\beta_\varepsilon^*(y_\varepsilon) \rightharpoonup \beta^*(y)$ in $L^2(Q_\delta)$. Then we get that $\zeta = \beta^*(y)$ a.e. on Q_δ and since δ is arbitrarily small we obtain $\zeta = \beta^*(y)$ a.e. on Q_u.

Here we have applied a consequence of Mazur theorem saying that if O is a bounded open set of finite measure and $(f_n)_{n\geq1}$ is a sequence bounded in $L^2(O)$ such that $f_n \to f$ a.e. on O, then $f_n \rightharpoonup f$ in $L^2(O)$ as $n \to \infty$.

The proof is continued as in Theorem 1.6 and Proposition 1.8.   □

1.3 Existence of Periodic Solutions in the Parabolic–Elliptic Degenerate Case

In this section we deal with the study of periodic solutions to the degenerate fast diffusion equation introduced in Sect. 1.1, under the hypothesis of a T-periodic function f. To this end, we first investigate the existence of a periodic solution to an intermediate problem restraint to a period T and extend then the result by periodicity to the time space $\mathbb{R}_+ = (0, \infty)$. The proof involves an appropriate approximating periodic problem and the existence of a solution is shown via a fixed point theorem on the basis of the results for the approximating problem (1.40). This result will also allow to characterize the behavior at large time of the solution to a Cauchy problem with periodic data.

We recall some previous papers dealing with periodic problems for degenerate linear equations. In [16] a problem of the type

$$\frac{d}{dt}(My(t)) + Ay(t) = f(t), \ 0 \leq t \leq 1,$$

with the periodic condition $(My)(0) = (My)(1)$ has been studied. Here M and A are two closed linear operators from a complex Banach space into itself, under the assumptions that the domain $D(A)$ of A is continuously embedded in $D(M)$ and A has a bounded inverse. Assuming suitable hypotheses on the modified resolvent $(\lambda M + A)^{-1}$, it has been proved that problem admits one 1-periodic solution. Some examples of applications to partial differential equations and ordinary differential equations have been given. The latter case has been studied in the paper [17].

The nondegenerate fast diffusion case with a nonlinear transport term has been approached in the paper [87], while the degenerate case without advection has been studied in [59].

As in Sect. 1.1, Ω is an open bounded subset of $\mathbb{R}^N$ and T is finite. We consider the problem

$$\frac{\partial(u(x)y)}{\partial t} - \Delta \beta^*(y) + \nabla \cdot K_0(x, y) \ni f \text{ in } \mathbb{R}_+ \times \Omega,$$

$$y(t, x) = 0 \text{ on } \mathbb{R}_+ \times \Gamma, \qquad (1.120)$$

$$(u(x)y(\tau, x))|_{\tau=t} - (u(x)y(\tau, x))|_{\tau=t+T} = 0 \text{ in } \mathbb{R}_+ \times \Omega,$$

under the assumption of the T-periodicity of the function f,

$$f(t, x) = f(t + T, x) \text{ a.e. } (t, x) \in \mathbb{R}_+ \times \Omega. \qquad (1.121)$$

The hypotheses made for β^*, K_0 and u are preserved as they were presented in Sect. 1.1 and we assume that $f \in L^\infty_{loc}(\mathbb{R}_+; V')$.

We begin with the study of the existence for the solution to the problem on a time period

$$\frac{\partial(u(x)y)}{\partial t} - \Delta \beta^*(y) + \nabla \cdot K_0(x, y) \ni f \text{ in } Q = (0, T) \times \Omega,$$

$$y(t, x) = 0 \text{ on } \Sigma = (0, T) \times \Gamma, \quad (1.122)$$

$$(u(x)y(t, x))|_{t=0} - (u(x)y(t, x))|_{t=T} = 0 \text{ in } \Omega.$$

Then, this solution will be extended by periodicity to all $t \in \mathbb{R}_+$.

1.3.1   Solution Existence on the Time Period $(0, T)$

The functional framework for this problem is the same as in Sect. 1.1.

Definition 1.11. Let $f \in L^\infty(0, T; V')$. We call a *weak solution* to (1.122) a pair (y, ζ) such that

$$y \in L^2(0, T; V), \ y(t, x) \leq y_s \text{ a.e. } (t, x) \in Q,$$

$$uy \in C([0, T]; L^2(\Omega)) \cap W^{1,2}([0, T]; V'),$$

$$\zeta \in L^2(0, T; V), \ \zeta(t, x) \in \beta^*(y(t, x)) \text{ a.e. } (t, x) \in Q,$$

satisfying the equation

$$\int_0^T \left\langle \frac{d(uy)}{dt}(t), \phi(t) \right\rangle_{V',V} dt + \int_Q (\nabla\zeta - K_0(x,y)) \cdot \nabla\phi dx dt$$
$$= \int_0^T \langle f(t), \phi(t) \rangle_{V',V} dt, \text{ for any } \phi \in L^2(0,T;V)$$

and the condition $(u(x)y(t,x))|_{t=0} - (u(x)y(t,x))|_{t=T} = 0$ in Ω.

With the same notation and definitions as in Sect. 1.1. we consider the periodic approximating problem

$$\frac{d(u_\varepsilon y_\varepsilon)}{dt}(t) + A_\varepsilon u_\varepsilon(t) = f(t) \text{ a.e. } t \in (0,T), \tag{1.123}$$

$$u_\varepsilon(y_\varepsilon(0) - y_\varepsilon(T)) = 0$$

which is equivalent with

$$\frac{d\theta_\varepsilon}{dt}(t) + B_\varepsilon \theta_\varepsilon(t) = f(t) \text{ a.e. } t \in (0,T), \tag{1.124}$$

$$\theta_\varepsilon(0) = \theta_\varepsilon(T),$$

by the function replacement $\theta_\varepsilon = u_\varepsilon y_\varepsilon$, with A_ε and B_ε given by (1.38) and (1.41), respectively.

Let us denote

$$C_f = \frac{2}{\rho} \left(\|f\|^2_{L^\infty(0,T;V')} + \overline{K}^2 \right), \tag{1.125}$$

where $\overline{K} = K_s(\text{meas}(\Omega))^{1/2} \sum_{j=1}^N a_j^M$ was defined in Proposition 1.5. We also recall that ρ was specified in (1.6), $\overline{M} = M_K \sum_{j=1}^N a_j^M$ and by c_P we have denoted the constant in the Poincaré inequality.

We are going to prove the following existence result.

Theorem 1.12. *Let $f \in L^\infty(0,T;V')$. Then, the periodic approximating problem (1.124) has a unique solution*

$$\theta_\varepsilon \in C([0,T];L^2(\Omega)) \cap W^{1,2}([0,T];V') \cap L^2(0,T;V), \tag{1.126}$$

$$\beta_\varepsilon^* \left(\frac{\theta_\varepsilon}{u_\varepsilon} \right) \in L^2(0,T;V). \tag{1.127}$$

Moreover, the solution satisfies the estimate

$$\int_0^T \left\| \frac{d\theta_\varepsilon}{d\tau}(\tau) \right\|_{V'}^2 d\tau + \int_0^T \left\| \beta_\varepsilon^* \left(\frac{\theta_\varepsilon}{u_\varepsilon}(\tau) \right) \right\|_V^2 d\tau \tag{1.128}$$

$$\leq 4 \left(\int_0^T \| f(t) \|_{V'}^2 dt + \overline{K}^2 T \right).$$

Proof. We apply a fixed point result and start this by fixing in (1.124) $\theta_\varepsilon(0)$ in $L^2(\Omega)$ and denoting it by v, i.e.,

$$\theta_\varepsilon(0) := v \in L^2(\Omega).$$

Hence we have to deal with the Cauchy problem

$$\frac{d\theta_\varepsilon}{dt}(t) + B_\varepsilon \theta_\varepsilon(t) = f(t) \text{ a.e. } t \in (0, T), \tag{1.129}$$

$$\theta_\varepsilon(0) = v,$$

whose well-posedness for $v \in L^2(\Omega)$ has already been studied in Sect. 1.1, Proposition 1.5. Thus, (1.129) has a unique solution (1.126)–(1.127).

Let us consider the set

$$\mathcal{S}_\varepsilon := \left\{ z \in L^2(\Omega); \; \left\| \frac{z}{\sqrt{u_\varepsilon}} \right\| \leq R_\varepsilon \text{ a.e. } x \in \Omega \right\} \tag{1.130}$$

where R_ε is a positive constant for each $\varepsilon > 0$. We define the mapping

$$\Psi_\varepsilon : \mathcal{S}_\varepsilon \to \mathcal{S}_\varepsilon, \; \Psi_\varepsilon(v) = \theta_\varepsilon(T), \text{ for any } v \in \mathcal{S}_\varepsilon$$

where $\theta_\varepsilon(t)$ is the solution to (1.129).

Since (1.129) has a unique solution for $v \in \mathcal{S}_\varepsilon$, the mapping Ψ_ε is single-valued and we are going to show that it has a fixed point by the Schauder–Tychonoff theorem (see e.g., [67], pp. 148), working in the weak topology. We begin by checking the conditions of this theorem.

(i1) It is obvious that $\mathcal{S}_\varepsilon$ is a convex, bounded and strongly closed subset of $L^2(\Omega)$. Hence it is weakly compact in $L^2(\Omega)$.

(i2) Next, we have to show the inclusion $\Psi_\varepsilon(\mathcal{S}_\varepsilon) \subset \mathcal{S}_\varepsilon$.

The solution $\theta_\varepsilon \in C([0, T]; L^2(\Omega))$ and so $\theta_\varepsilon(T) = u_\varepsilon y_\varepsilon(T) \in L^2(\Omega)$. We test (1.129) for $\frac{\theta_\varepsilon}{u_\varepsilon} \in V$ and recalling (1.37) and (1.14) we get

$$\frac{1}{2} \frac{d}{dt} \left\| \frac{\theta_\varepsilon}{\sqrt{u_\varepsilon}}(t) \right\|^2 + \rho \left\| \frac{\theta_\varepsilon}{u_\varepsilon}(t) \right\|_V^2 \leq \| f(t) \|_{V'} \left\| \frac{\theta_\varepsilon}{u_\varepsilon}(t) \right\|_V + \overline{K} \left\| \frac{\theta_\varepsilon}{u_\varepsilon}(t) \right\|_V$$

$$\leq \frac{\rho}{2} \left\| \frac{\theta_\varepsilon}{u_\varepsilon}(t) \right\|_V^2 + \frac{1}{\rho} \left(\| f \|_{L^\infty(0,T;V')}^2 + \overline{K}^2 \right).$$

Next, applying the Poincaré inequality we have

$$\frac{d}{dt}\left\|\frac{\theta_\varepsilon}{\sqrt{u_\varepsilon}}(t)\right\|^2 + \frac{\rho}{c_P^2}\left\|\frac{\theta_\varepsilon}{u_\varepsilon}(t)\right\|^2 \leq C_f$$

and using the relation $u_\varepsilon(x) \leq u_M + \varepsilon < u_M + 1$ (since ε is arbitrarily small) we obtain

$$\frac{d}{dt}\left\|\frac{\theta_\varepsilon}{\sqrt{u_\varepsilon}}(t)\right\|^2 + \rho_0\left\|\frac{\theta_\varepsilon}{\sqrt{u_\varepsilon}}(t)\right\|^2 \leq C_f$$

with $\rho_0 = \frac{\rho}{(u_M+1)c_P^2}$. Integrating on $(0,t)$ with $t \in [0,T]$ we get

$$\left\|\frac{\theta_\varepsilon}{\sqrt{u_\varepsilon}}(t)\right\|^2 \leq \left\|\frac{v}{\sqrt{u_\varepsilon}}\right\|^2 \exp(-\rho_0 t) + \frac{C_f}{\rho_0}(1 - \exp(-\rho_0 t)).$$

Now if $R_\varepsilon^2 \geq \frac{C_f}{\rho_0}$ (and this is true since R_ε is large enough) and $v \in \mathcal{S}_\varepsilon$ it follows that

$$\left\|\frac{\theta_\varepsilon}{\sqrt{u_\varepsilon}}(t)\right\| \leq R_\varepsilon, \text{ for any } t \in [0,T].$$

Thus, we have obtained that $\theta_\varepsilon(T) = \Psi_\varepsilon(v) \in \mathcal{S}_\varepsilon$ and therefore, it follows that $\Psi_\varepsilon(\mathcal{S}_\varepsilon)$ is weakly compact, too.

(i3) Finally, we have to prove that the mapping Ψ_ε is weakly continuous.

For that we consider a sequence

$$\{v^n\}_{n\geq 1} \subset \mathcal{S}_\varepsilon, \; v^n \rightharpoonup v \text{ in } L^2(\Omega) \text{ as } n \to \infty,$$

and will show that

$$\Psi_\varepsilon(v^n) \rightharpoonup \Psi_\varepsilon(v) \text{ in } L^2(\Omega) \text{ as } n \to \infty.$$

We introduce the approximating problem

$$\frac{d\theta_\varepsilon^n}{dt}(t) + B_\varepsilon\theta_\varepsilon^n(t) = f(t), \text{ a.e. } t \in (0,T),$$

$$\theta_\varepsilon^n(0) = v^n.$$

This has a unique solution

$$\theta_\varepsilon^n \in C([0,T];V') \cap W^{1,2}([0,T];V') \cap L^2(0,T;V), \; \beta_\varepsilon^*\left(\frac{\theta_\varepsilon^n}{u_\varepsilon}\right) \in L^2(0,T;V)$$

satisfying the estimate (1.52). Now, by (1.59)

$$\int_\Omega u_\varepsilon j_\varepsilon \left(\frac{\theta_\varepsilon^n}{u_\varepsilon}\right) dx \leq (u_M + \varepsilon) \frac{\beta_s^* - \beta^*(y_s - \varepsilon)}{2\varepsilon} \left\|\frac{v^n}{u_\varepsilon}\right\|^2$$

and

$$\left\|\frac{v^n}{u_\varepsilon}\right\| \leq \frac{1}{\sqrt{\varepsilon}} \left\|\frac{v^n}{\sqrt{u_\varepsilon}}\right\| \leq \frac{1}{\sqrt{\varepsilon}} R_\varepsilon$$

due to the fact that $v^n \in \mathcal{S}_\varepsilon$. Therefore (1.52) written for θ_ε^n is bounded independently of n, and we can proceed like in Proposition 1.5 to show that θ_ε^n tends in some appropriate space to θ_ε which turns out to be the solution to (1.129). This implies also the convergence

$$\theta_\varepsilon^n(T) \to \theta_\varepsilon(T) \text{ in } V', \text{ as } n \to \infty$$

due to the Ascoli–Arzelà theorem (see (1.62)). Hence

$$\Psi_\varepsilon(v^n) = \theta_\varepsilon^n(T) \rightharpoonup \theta_\varepsilon(T) = \Psi_\varepsilon(v) \text{ in } L^2(\Omega),$$

and because $\mathcal{S}_\varepsilon$ is weakly closed it follows that $\theta_\varepsilon(T) \in \mathcal{S}_\varepsilon$.

Now the Schauder–Tychonoff theorem ensures that Ψ_ε has a fixed point, implying that

$$\theta_\varepsilon(0) = \theta_\varepsilon(T) \text{ or } u_\varepsilon \theta_\varepsilon(0) = u_\varepsilon \theta_\varepsilon(T).$$

Consequently, (1.124) has at least a solution.

The estimate (1.128) follows immediately by (1.52) in Proposition 1.5, for $t = T$.

Uniqueness is proved as in Proposition 1.5, taking the same data in (1.53). This ends the proof of Theorem 1.12. $\qquad\square$

Theorem 1.13. *Let $f \in L^\infty(0, T; V')$. Then, the periodic problem (1.122) has at least a solution (y^*, ζ) such that*

$$y^* \in L^2(0, T; V),$$

$$uy^* \in C([0, T]; L^2(\Omega)) \cap W^{1,2}([0, T]; V'),$$

$$\zeta \in L^2(0, T; V), \ \zeta(t, x) \in \beta^*(y^*(t, x)) \text{ a.e. } (t, x) \in Q,$$

$$y^*(t, x) \leq y_s \text{ a.e. } (t, x) \in Q.$$

If (1.107) and

$$\rho > N M_K k_u c_P \sqrt{u_M} \tag{1.131}$$

are satisfied the solution is unique a.e. on Q.

Proof. The proof of the existence is based on the same arguments and is led in the same way as in Theorem 1.6, including the construction of y^*, with the corresponding modifications due to the periodicity condition. Thus in the approximating problem in Theorem 1.6, $\theta_\varepsilon(0) = u_\varepsilon y_\varepsilon(0) = u_\varepsilon y_\varepsilon(T) = \theta_\varepsilon(T)$ and by (1.80) we get $(uy)|_{t=0} = (uy)|_{t=T}$ property which is inherited by uy^*. Obviously, $uy^* = 0$ in Q_0.

Assume now (1.107) and that there exist two solutions (y^*, ζ) and $(\overline{y}^*, \overline{\zeta})$ to (1.122) corresponding to the same periodic data f. We subtract (1.122) written for y^* and $\overline{y}^*$ and multiply the difference scalarly in V' by $u(y^* - \overline{y}^*)(t)$,

$$\left(\frac{d(u(y^* - \overline{y}^*))}{dt}(t), u(y^* - \overline{y}^*)(t) \right)_{V'} + \int_\Omega \nabla(\zeta(t) - \overline{\zeta}(t)) \cdot \nabla \psi(t) dx$$

$$= \int_\Omega (K(y^*(t)) - K(\overline{y}^*(t)) a(x) \cdot \nabla \psi(t) dx$$

where $A_0 \psi(t) = u(y^* - \overline{y}^*)(t)$, a.e. t, (where we recall that $A_0 = -\Delta$ with Dirichlet boundary conditions (see (1.17)). Integrating over $(0, T)$ and proceeding as in Proposition 1.8 we get

$$\frac{1}{2} \|u(y^* - \overline{y}^*)(T)\|_{V'}^2 - \frac{1}{2} \|u(y^* - \overline{y}^*)(0)\|_{V'}^2 + \frac{\rho}{2} \int_0^T \int_\Omega u(y^* - \overline{y}^*)^2 dxdt$$

$$\leq \frac{1}{2\rho} (N M_K k_u)^2 \int_0^T \|u(y^* - \overline{y}^*)(\tau)\|_{V'}^2 d\tau$$

$$\leq \frac{1}{2\rho} (N M_K k_u c_P \sqrt{u_M})^2 \int_0^T \left\| \sqrt{u}(y^* - \overline{y}^*)(\tau) \right\|^2 d\tau,$$

where c_P is the constant in the Poincaré inequality. We apply the solution periodicity and it remains that $\|\sqrt{u}(y^* - \overline{y}^*)\|_{L^2(Q)}^2 = 0$. This implies that $uy^* = u\overline{y}^*$ a.e. on Q and then we continue as in Proposition 1.8. $\square$

1.3.2 Solution Existence on $\mathbb{R}_+$

Now we can extend the previous result to $t \in \mathbb{R}_+$. We resume problem (1.120) and prove

Theorem 1.14. *Let us assume*

$$f \in L_{loc}^\infty(\mathbb{R}_+; V'), \ f(t, x) = f(t + T, x) \ a.e. \ (t, x) \in \mathbb{R}_+ \times \Omega.$$

Then problem (1.120) has at least a solution $y \in L^2_{loc}(\mathbb{R}_+; V)$ *satisfying*

$$\theta = uy \in C(\mathbb{R}_+; L^2(\Omega)) \cap W^{1,2}_{loc}(\mathbb{R}_+; V'),$$

$$y(t, x) \leq y_s \text{ a.e. } (t, x) \in \mathbb{R}_+ \times \Omega,$$

$$\zeta \in L^2_{loc}(\mathbb{R}_+; V), \text{ where } \zeta(t, x) \in \beta^*(y(t, x)) \text{ a.e. } (t, x) \in \mathbb{R}_+ \times \Omega.$$

If (1.107) and (1.131) are satisfied then the solution is unique.

Proof. We consider first (1.120) on $(0, T)$ with $f|_{(0,T)}$. We obtain (1.122) which has a periodic solution with $(u(x)y(t, x))|_{t=0} = (u(x)y(t, x))|_{t=T}$ in Ω. Then we consider (1.120) on $(T, 2T)$ with the periodicity condition $(u(x)y(t, x))|_{t=T} = (u(x)y(t, x))|_{t=2T}$ in Ω. We make the transformation $t' = t - T$ and denote $\widetilde{y}(t', x) = y(t' + T, x)$ with $t' \in [0, T]$. Using now the periodicity of the function f we find again problem (1.122) which has a periodic solution $\widetilde{y}(t')$ with $\widetilde{\theta} = u\widetilde{y} \in C([0, T]; L^2(\Omega))$, such that $(u(x)\widetilde{y}(t', x))|_{t'=0} = (u(x)\widetilde{y}(t', x))|_{t'=T}$. Coming back to the variable t we obtain that (1.120) has a periodic solution such that $u(x)y(t, x)$ is continuous on $[T, 2T]$ and this extends by continuity the solution obtained on $[0, T]$. The procedure is continued in this way on each time period. Moreover, if a satisfies (1.107) and (1.131) the solution is unique on each period. $\square$

1.3.3   Longtime Behavior of the Solution to a Cauchy Problem with Periodic Data

Finally we are going to characterize the longtime behavior of the solution y to problem (1.1) with a T-periodic function f. The domain Q is taken $\mathbb{R}_+ \times \Omega$, and we assume that the solution starts from the initial condition θ_0. Let

$$f \in L^\infty_{loc}(\mathbb{R}_+; V'), \quad f(t, x) = f(t + T, x) \text{ a.e. } (t, x) \in \mathbb{R}_+ \times \Omega, \qquad (1.132)$$

$$\theta_0 \in L^2(\Omega), \quad \theta_0 = 0 \text{ a.e. on } \Omega_0,$$

$$\theta_0 \geq 0 \text{ a.e. on } \Omega_u, \quad \frac{\theta_0}{u} \in L^2(\Omega_u), \quad \frac{\theta_0}{u}(x) \leq y_s \text{ a.e. } x \in \Omega_u$$

and we recall that u_M is the maximum of u and c_P is the constant in Poincaré inequality (1.19).

Proposition 1.15. *Let us assume (1.107) and (1.131). Then, the solution to the Cauchy problem (1.1) with f periodic of period T satisfies*

$$\lim_{t \to \infty} \|(uy - u\omega)(t)\|_{V'} = 0 \qquad (1.133)$$

exponentially, where ω is the unique periodic solution to (1.120) and y is the unique solution to (1.1).

Proof. By Theorem 1.14 the solution to (1.120) is unique and let us denote it by ω. We multiply the difference of (1.1) and (1.120) by $u(y(t) - \omega(t))$ scalarly in V', and we get

$$\frac{1}{2}\frac{d}{dt}\|u(y-\omega)(t)\|_{V'}^2 + \rho \int_\Omega u(y-\omega)^2(t)dx$$

$$\leq NM_K k_u c_P \sqrt{u_M}\left\|\sqrt{u}(y-\omega)(t)\right\|^2.$$

Therefore, applying (1.131) we obtain

$$\frac{d}{dt}\|u(y-\omega)(t)\|_{V'}^2 + \delta\left\|\sqrt{u}(y-\omega)(t)\right\|^2 \leq 0$$

with $\delta = \rho - NM_K k_u c_P \sqrt{u_M}$.

We have that

$$\int_\Omega u(y(t) - \omega(t))^2 dx \geq \frac{1}{u_M}\int_\Omega u^2(y(t) - \omega(t))^2 dx \geq \frac{1}{u_M c_P^2}\|u(y-\omega)(t)\|_{V'}^2,$$

hence

$$\frac{d}{dt}\|u(y-\omega)(t)\|_{V'}^2 + \delta_0\|u(y-\omega)(t)\|_{V'}^2 \leq 0$$

with $\delta_0 = \frac{\delta}{u_M c_P^2}$. We deduce that

$$\|u(y-\omega)(t)\|_{V'}^2 \leq e^{-\delta_0 t}\|\theta_0 - (u\omega)(0)\|_{V'}^2$$

and this implies (1.133).   □

Referring to applications in real-world problems we remark that the behavior (1.133) of the solution to the Cauchy problem (1.1) with a periodic f is possible only if the advection is done with a velocity in absolute value lower than the porosity u and the diffusion processes has a sufficient high diffusion coefficient. This means that the velocity must be sufficient small in comparison with the pore dimension and that the diffusivity should dominate the advection.

1.3.4   Numerical Results

We shall provide some simulations intended to show the behavior at large time of the solution to (1.1) with a periodic f.

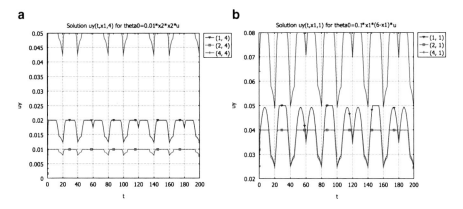

Fig. 1.6 Asymptotic behavior of $\theta = uy$ solution to (1.1) in the periodic parabolic–elliptic degenerate case

The computations are done in 2D with $\Omega = (0,5) \times (0,5)$, with the same data for Ω_0, u, β^* and K as in Sect. 1.1, (1.110), (1.109), $a = (1,1)$, $c = 1.5$ (a weakly nonlinear porous medium),

$$f(t,x_1,x_2) = \begin{cases} \left(\left|\sin \frac{\pi}{20}t\right| + \left|\cos \frac{\pi}{30}t\right|\right), & x \in \Omega_u \\ 0, & x \in \Omega_0 \end{cases}$$

and two different initial data. In Fig. 1.6a the values $\theta(t,x) = u(x)y(t,x)$ are computed for

$$\theta_0(x_1,x_2) = 0.01x_2^2u(x_1,x_2)$$

and represented at $x = (x_1, 4)$, $x_1 = 1, 2, 4$.

In Fig. 1.6b there are the graphics $\theta(t,x) = u(x)y(t,x)$ at $x = (x_1,1)$, $x_1 = 1, 2, 4$, computed for

$$\theta_0(x_1,x_2) = 0.1x_1(6 - x_1)u(x_1,x_2). \tag{1.134}$$

We can see that after some time the solutions to (1.1) become periodic.

Chapter 2
Existence for Diffusion Degenerate Problems

In this chapter we present another method for studying the well-posedness of a multivalued degenerate fast diffusion equation by proposing an appropriate time discretization scheme. We consider that the degeneration of the equation is due to the vanishing of the diffusion coefficient and choose for this problem Robin boundary conditions which contain the multivalued function as well.

In this case the operator defined in the abstract formulation of the problem is not strongly monotone (and not invertible) because of the lack of strictly monotonicity of the function β^*. Therefore, arguments following the general theory (see [43]), related to the existence of a mild solution to a Cauchy problem with an m-accretive operator on a Banach space (see also [11,45,73]) cannot be applied directly. We introduce a time-discretization scheme with a quasi m-accretive operator and develop a direct proof of the stability and convergence of it. This also enables the achievement of more precise results concerning the convergence of the discretized solution to the solution to the original problem. Besides the proof of the solution existence, this approach is aimed to be a mathematical background to sustain the correctness of the numerical algorithm for computing the solution to this type of equations by avoiding the approximation of the multivalued function. The method was developed for the nondegenerate singular fast diffusion case in [39] and treated in the paper [88] for the degenerate situation which will be presented here.

2.1 Well-Posedness for the Cauchy Problem with Fast Diffusion

As usually we consider Ω an open bounded subset of $\mathbb{R}^N$ ($N \in \mathbb{N}^* = \{1, 2, \ldots\}$), with the boundary $\Gamma := \partial\Omega$ sufficiently smooth. We shall deal with the boundary value problem with initial data

A. Favini and G. Marinoschi, *Degenerate Nonlinear Diffusion Equations*, Lecture Notes in Mathematics 2049, DOI 10.1007/978-3-642-28285-0_2, © Springer-Verlag Berlin Heidelberg 2012

$$\frac{\partial y}{\partial t} - \Delta \beta^*(y) + \nabla \cdot K_0(x,y) \ni f \text{ in } Q := (0,T) \times \Omega, \qquad (2.1)$$

$$y(0,x) = y_0 \text{ in } \Omega \qquad (2.2)$$

and illustrate the theory by considering Robin boundary conditions of the form

$$(K_0(x,y) - \nabla \beta^*(y)) \cdot \nu - \alpha \beta^*(y) \ni 0 \text{ on } \Sigma := (0,T) \times \Gamma, \qquad (2.3)$$

where ν is the unit outward normal vector to Γ.

In this model β^* is the multivalued function introduced in Sect. 1.1 having the properties specified there, except for (1.3) which is assumed here to take place with $\rho = 0$, i.e.,

$$\beta(r) \geq \gamma_\beta |r|^m \text{ for any } r \leq 0, \qquad (2.4)$$

with $m > 0$ and $\gamma_\beta > 0$.

This implies that (1.7) is replaced by

$$(\zeta - \overline{\zeta})(r - \overline{r}) \geq 0, \ \forall r, \overline{r} \in (-\infty, y_s], \ \zeta \in \beta^*(r), \ \overline{\zeta} \in \beta^*(\overline{r}), \qquad (2.5)$$

so that in this case the degeneracy of the equation is induced by this property.

In a physical model the function α characterizes the space variable permeability of the boundary Γ. The boundary condition (2.3) expresses the fact that the flux across the boundary is proportional to the diffusivity and to the variable permeability of the boundary Γ and no subset of it is impermeable.

Taking into account the properties of β^* it follows that $(\beta^*)^{-1} : \mathbb{R} \to (-\infty, y_s]$ is single valued, continuous and monotonically increasing on $(-\infty, \beta_s^*)$ and constant on $[\beta_s^*, +\infty)$

$$(\beta^*)^{-1}(r) \leq y_s \text{ for } r \leq \beta_s^*, \ (\beta^*)^{-1}(r) = y_s \text{ for } r > \beta_s^*. \qquad (2.6)$$

We shall assume that $K_0 : \Omega \times (-\infty, y_s]$, having the same meaning as in Chap. 1, is linear with respect to the second variable

$$K_0(x,r) = a_i(x)r, \ i = 1, \ldots, N, \qquad (2.7)$$

namely that the advection has the velocity $a(x)$ and is supposed to depend linearly on the solution. From the mathematical point of view such a requirement is necessary in all results related to existence, Proposition 2.5, estimate (2.36), Theorem 2.6 (and will be explained at the end of the proof of Theorem 2.6). Here we still assume that

$$a_i \in W^{1,\infty}(\Omega), \qquad (2.8)$$

satisfying

$$a \cdot \nu \leq 0 \text{ on } \Gamma. \tag{2.9}$$

In this section we assume that

$$f \in C([0,T]; L^2(\Omega)), \tag{2.10}$$

and that $\alpha : \Gamma \to [\alpha_m, \alpha_M]$ is continuous and positive,

$$\alpha(x) \geq \alpha_m := \min_{x \in \Gamma} \alpha(x) > 0 \text{ on } \Gamma. \tag{2.11}$$

The definition of the function j given by (1.32) and its properties remain valid.

2.1.1 Functional Framework and Time Discretization Scheme

As done before we denote the scalar product and norm in $L^2(\Omega)$ with no subscript, i.e., by $(\cdot, \cdot)$ and $\|\cdot\|$, respectively.

We have to work in the space $V = H^1(\Omega)$, choosing the norm

$$\|\psi\|_V = \left(\int_\Omega |\nabla \psi(x)|^2 \, dx + \int_\Gamma \alpha(x) |\psi(x)|^2 \, d\sigma \right)^{1/2}, \tag{2.12}$$

which is equivalent with the standard Hilbertian norm on the space $H^1(\Omega)$, denoted by $\|\cdot\|_{H^1(\Omega)}$. By the inequality

$$\|\psi\|^2_{H^1(\Omega)} \leq c_P \left(\int_\Omega |\nabla \psi(x)|^2 \, dx + \int_\Gamma |\psi(x)|^2 \, d\sigma \right), \quad \forall \psi \in V, \tag{2.13}$$

(see [91], pp. 20), and the trace theorem (see [13], pp. 122) we deduce that there exist positive constants denoted $c_V, c_H, c_{\Gamma_\alpha}$ depending on $\alpha_m^{-1/2}$, the domain Ω and the dimension N, such that for any $\psi \in V$ we have

$$\|\psi\|_{H^1(\Omega)} \leq c_H \|\psi\|_V, \quad \|\psi\|_V \leq c_V \|\psi\|_{H^1(\Omega)}, \quad \|\psi\|_{L^2(\Gamma)} \leq c_{\Gamma_\alpha} \|\psi\|_V. \tag{2.14}$$

The scalar product in the dual space V' is introduced by

$$(y, \overline{y})_{V'} = \langle y, A_\Delta^{-1} \overline{y} \rangle_{V',V}, \quad \text{for any } y, \overline{y} \in V', \tag{2.15}$$

where $\langle \cdot, \cdot \rangle_{V',V}$ is the pairing between V' and V, and $A_\Delta : V \to V'$ is the operator defined by

$$\langle A_\Delta \psi, \phi \rangle_{V',V} = \int_\Omega \nabla \psi \cdot \nabla \phi dx + \int_\Gamma \alpha(x) \psi \phi d\sigma, \text{ for any } \psi, \ \phi \in V. \quad (2.16)$$

Definition 2.1. Let $f \in C([0,T]; L^2(\Omega))$ and

$$y_0 \in L^2(\Omega), \ y_0 \leq y_s \text{ a.e. } x \in \Omega$$

hold. A *weak solution* to (2.1)–(2.3) is a pair (y, ζ),

$$y \in L^2(0,T; L^2(\Omega)) \cap W^{1,2}([0,T]; V'),$$

$$\zeta \in L^2(0,T;V), \ \zeta(t,x) \in \beta^*(y(t,x)) \text{ a.e. } (t,x) \in Q,$$

which satisfies the equation

$$\left\langle \frac{dy}{dt}(t), \psi \right\rangle_{V',V} + \int_\Omega (\nabla \zeta(t) - K_0(x, y(t))) \cdot \nabla \psi dx \quad (2.17)$$

$$= \int_\Omega f(t)\psi dx - \int_\Gamma \alpha \zeta(t) \psi d\sigma, \text{ a.e. } t \in (0,T), \text{ for any } \psi \in V,$$

along with the initial and boundedness conditions

$$y(0,x) = y_0 \text{ in } \Omega,$$

$$y(t,x) \leq y_s \text{ a.e. } (t,x) \in Q.$$

An equivalent form to (2.17) is

$$\int_0^T \left\langle \frac{dy}{dt}(t), \phi \right\rangle_{V',V} dt + \int_Q (\nabla \zeta - K_0(x,y)) \cdot \nabla \phi dx dt$$

$$= \int_0^T \int_\Omega f\psi dx dt - \int_\Sigma \alpha \zeta \phi d\sigma dt, \text{ a.e. } t \in (0,T), \quad (2.18)$$

for any $\phi \in L^2(0,T;V)$.

We remark that the weak solution in the sense of Definition 2.1 satisfies (2.1) in the sense of distributions. Concerning (2.3), at a first glance this might occur as not correct because the solution y does not have a trace on Σ. We specify that this way of writing is formal and the rigorous interpretation is that $(y(t), \zeta(t))$, the weak solution to (2.1)–(2.3), is the limit of a sequence $(y^h(t), \zeta^h(t))$, with $y^h(t)$, $\zeta^h(t)$ belonging to V. Thus, the flux $(a(x)K(y^h(t)) - \nabla \beta^*(y^h(t)) \in V$ and his trace on Γ is well defined a.e. t in the sense of (1.93). This will be in fact the result proved in this section.

We introduce the multivalued operator $A : D(A) \subset V' \to V'$ with the domain

$$D(A) = \{y \in L^2(\Omega); \text{ there exists } \zeta \in V, \ \zeta(x) \in \beta^*(y(x)) \text{ a.e. } x \in \Omega\},$$

and define it by the relation

$$\langle Ay, \psi \rangle_{V',V} = \int_\Omega (\nabla \zeta - K_0(x,y)) \cdot \nabla \psi dx + \int_\Gamma \alpha \zeta \psi d\sigma, \text{ for any } \psi \in V, \tag{2.19}$$

and $\zeta(x) \in \beta^*(y(x))$ a.e. $x \in \Omega$.

With this notation we can introduce the abstract Cauchy problem

$$\frac{dy}{dt}(t) + Ay(t) \ni f(t), \text{ a.e. } t \in (0,T), \tag{2.20}$$

$$y(0) = y_0. \tag{2.21}$$

The well-posedness of the problem (2.20)–(2.21) can be investigated in a direct way, as done in Sect. 1.1, by using a regular approximation β_ε^* of β^*, and proving the existence and convergence of the approximating solution to a solution to (2.20)–(2.21). Even if this method of approximating β^* is efficient to prove the existence of solutions to these problems, numerical simulations performed according to it fail for ε small due to the fact that β_ε (the corresponding approximation of β) blows up as $\varepsilon \to 0$. To overpass this inconvenience and to set a background for a numerical algorithm we shall study a scheme with time differences which is efficient for this purpose.

First let us introduce another variant of the concept of a *mild* solution.

Let h be positive. A *h-discretization* on $[0,T]$ for (2.20) consists in a partition $0 = t_0 \leq t_1 \leq t_2 \leq \ldots \leq t_n$ of the interval $[0,T]$, with $t_i - t_{i-1} = h$ for $i = 1, \ldots, n$, and a finite sequence $(f_i^h)_{i=1,\ldots,n}$, such that there exists $\delta(h)$ which tends to 0 as $h \to 0$ and

$$\|f(t) - f_i^h\| \leq \delta(h), \ t \in (t_{i-1}, t_i).$$

The h-discretization is denoted by $D_A^h(0 = t_0 \leq t_1 \leq t_2 \leq \ldots \leq t_n; f_1^h, \ldots, f_n^h)$. The time step $h = \frac{T}{n}$ and n will be further determined.

Here we compute f_i^h as the time averages

$$f_i^h = \frac{1}{h} \int_{(i-1)h}^{ih} f(s)ds \tag{2.22}$$

and see by (2.10) that $f_i^h \in L^2(\Omega)$.

We propose as time discretized system

$$\left(\frac{1}{h}I + A^h\right) y_i^h \ni f_i^h + \frac{1}{h}y_{i-1}^h, \ i = 1, \ldots, n, \tag{2.23}$$

and set

$$y_0^h = y_0 \quad \text{in} \quad \Omega. \tag{2.24}$$

The operator $A^h : D(A^h) \subset V' \to V'$ is multivalued, defined by

$$D(A^h) = \{y \in V; \text{ there exists } \zeta \in V, \ \zeta(x) \in \beta^*(y(x)) \text{ a.e. } x \in \Omega\},$$

$$\langle A^h y, \psi \rangle_{V',V} = \int_\Omega \left(\nabla \zeta + \sqrt{h} \nabla y - a(x)y \right) \cdot \nabla \psi dx + \int_\Gamma \alpha(\zeta + \sqrt{h}y)\psi d\sigma,$$

for any $\psi \in V$.

Definition 2.2. Let $y_0 \in L^2(\Omega)$, $y_0 \leq y_s$ a.e. $x \in \Omega$. A *h-approximate solution* to (2.20)–(2.21) in relation with the h-discretization $D_A^h (0 = t_0 \leq t_1 \leq t_2 \leq \ldots \leq t_n; f_1^h, \ldots, f_n^h)$ is a piecewise constant function denoted $y^h :$ $[0,T] \to V'$ whose restrictions y_i^h on $(t_{i-1}, t_i]$ satisfy (2.23) with $y^h(0) = y_0^h$.
  We say that y is a *mild solution* to (2.20)–(2.21) if $y \in C([0,T]; V')$ and for each h there exists a h-approximate solution z , such that

$$\|y(t) - z(t)\|_{V'} \leq \varrho(h) \text{ for all } t \in [0, T]$$

and $y(0) = y_0$, where ϱ is continuous and $\varrho(0) = 0$.

As a matter of fact, by the method further developed we shall prove the existence of a *mild solution* to (2.20)–(2.21) and show that is has the supplementary properties specified in Definition 2.1.

We are concerned with the stability and the convergence of the scheme (2.23)–(2.24), emphasizing the precise nature of its convergence towards the solution to (2.20)–(2.21). At the same time we shall show that a mild solution to the abstract Cauchy problem (2.20)–(2.21) is in fact a weak solution to (2.1)–(2.3).

The method allows to compute the solution without approximating the inclusion β^* and this enables the construction of an algorithm for the numerical computation of the solution in the subsets where β^* is multivalued.

2.1.2 Stability of the Discretization Scheme

Equation (2.23) can be written in the equivalent form

$$\left\langle \frac{y_i^h - y_{i-1}^h}{h}, \psi \right\rangle_{V',V} + \int_\Omega (\nabla \zeta_i^h + \sqrt{h} \nabla y_i^h - a(x)y_i^h) \cdot \nabla \psi dx \tag{2.25}$$

$$+ \int_\Gamma \alpha(x)(\zeta_i^h + \sqrt{h}y_i^h)\psi d\sigma = \int_\Omega f_i^h \psi dx,$$

for any $\psi \in V$, where $\zeta_i^h(x) \in \beta^*(y_i^h(x))$ a.e. $x \in \Omega$ and the first result is that it has a unique solution, y_i^h belonging to $D(A^h)$, for each $i = 1, \ldots, n$.

The existence for (2.23) with h fixed, small enough, follows by the quasi m-accretivity of the operator A^h.

Proposition 2.3. *The operator A^h is quasi m-accretive.*

Proof. The computations are similar to those done in Lemma 1.4, so that we shall not give all details. For the quasi accretivity let $\lambda > 0$, $\zeta \in A^h y$, $\overline{\zeta} \in A^h \overline{y}$ a.e. on Ω, and compute, according to (2.15)

$$\left(\lambda(y - \overline{y}) + \zeta - \overline{\zeta}, y - \overline{y} \right)_{V'} = \lambda \|y - \overline{y}\|_{V'}^2 + \int_\Omega \left(\zeta(y) - \zeta(\overline{y}) \right) (y - \overline{y}) \, dx$$

$$+ \sqrt{h} \int_\Omega (y - \overline{y})^2 \, dx - a_0 \|y - \overline{y}\| \, \|y - \overline{y}\|_{V'} \, ,$$

where

$$a_0 = \sum_{i=1}^N \|a_i\|_{L^\infty(\Omega)} \, . \tag{2.26}$$

Therefore, using (2.5) we obtain

$$\left(\lambda(y - \overline{y}) + \zeta - \overline{\zeta}, y - \overline{y} \right)_{V'} \geq \left(1 - \frac{1}{k} \right) \sqrt{h} \|y - \overline{y}\|^2 + \left(\lambda - \frac{a_0^2 k}{\sqrt{h}} \right) \|y - \overline{y}\|_{V'}^2 \tag{2.27}$$

which is nonnegative for $\lambda \geq \lambda_0 = \dfrac{a_0^2 k}{\sqrt{h}}$, where k is an arbitrary number, $k > 1$.

For the quasi m-accretivity we have to show that $R(\lambda I + A^h) = V'$ for $\lambda > \lambda_0$, i.e., to find for each $v \in V'$ a solution $y \in D(A^h)$ to the equation

$$\left(\lambda I + A^h \right) y = v. \tag{2.28}$$

Let us denote

$$\widetilde{\beta}^*(y) = \beta^*(y) + \sqrt{h} y \tag{2.29}$$

and deduce by the properties of β^* that

$$(\zeta - \overline{\zeta})(y - \overline{y}) \geq \sqrt{h}(y - \overline{y})^2, \ y, \overline{y} \in (-\infty, y_s],$$

where $\zeta \in \widetilde{\beta}^*(y)$, $\overline{\zeta} \in \widetilde{\beta}^*(\overline{y})$ a.e. on Ω. Let $\zeta \in \widetilde{\beta}^*(y)$. Its inverse $G^h \zeta = \left(\widetilde{\beta}^* \right)^{-1}(\zeta)$ is Lipschitz on $L^2(\Omega)$ with the constant $\frac{1}{\sqrt{h}}$ and continuous from V to $L^2(\Omega)$. Namely we have

$$\left\| G^h \zeta - G^h \overline{\zeta} \right\| \leq \frac{1}{\sqrt{h}} \left\| \zeta - \overline{\zeta} \right\| \leq \frac{c_H}{\sqrt{h}} \left\| \zeta - \overline{\zeta} \right\|_V \, ,$$

with c_H specified in (2.14). We can rewrite (2.28) as

$$\lambda G^h \zeta + A_1 \zeta = v, \tag{2.30}$$

where $A_1 : V \to V'$ is defined by

$$\langle A_1\zeta, \psi\rangle_{V',V} = \int_\Omega \nabla\zeta \cdot \nabla\psi dx - \int_\Omega a(x)G^h\zeta \cdot \nabla\psi dx + \int_\Gamma \alpha(x)\zeta\psi d\sigma, \ \forall\psi \in V$$

and show that the operator $B^h = \lambda G^h + A_1 : V \to V'$ is monotone, continuous and coercive for λ large enough. The proof follows like in Lemma 1.4 in Sect. 1.1. Therefore, B^h is surjective and (2.30) has a solution $\zeta \in V$, $\zeta \in \tilde{\beta}^*(y)$ a.e. in Ω, implying that $y \in V$ (because the inverse of $\tilde{\beta}^*$ is Lipschitz). In fact we have proved that $y \in D(A^h)$ is a solution to (2.28). $\qquad\square$

Consequently, we can notice that if we take $\lambda = \frac{1}{h}$ in (2.27), then for

$$h < \frac{1}{k^2}\frac{1}{a_0^4} \tag{2.31}$$

the operator $\frac{1}{h}I + A^h$ is invertible. Relationship (2.31) is a first estimate for the maximum time step. Therefore, the number n can be chosen such that $n \geq \left[k^2 a_0^4 T\right] + 1$.

We give now some preliminary results. First we notice that by (2.5) we obtain by a straightforward computation

$$\int_\Omega \nabla y_i^h \cdot \nabla\zeta_i^h dx \geq 0, \quad \int_\Gamma \alpha(x)y_i^h\zeta_i^h d\sigma \geq 0. \tag{2.32}$$

Next, we recall the following result, which stands for a discrete version of Gronwall's lemma (see this proof in [39]).

Further we take into account (2.31) and the relation above connecting n and T.

Lemma 2.4. *Let $v_i \in L^2(\Omega)$, $i = 0, \ldots, n$, such that*

$$\|v_p\|^2 \leq C_M h \sum_{i=1}^{p} \|v_i\|^2 + C_0, \ p = 1, \ldots, n, \tag{2.33}$$

where $C_M > 0$, $C_0 \geq 0$. Then

$$\|v_p\|^2 \leq 2\max\{1, C_M\}e^{C_M T}(\|v_0\|^2 + C_0) \tag{2.34}$$

and

$$h\sum_{i=1}^{p} \|v_i\|^2 \leq \max\left\{1, \frac{1}{C_M}\right\}e^{C_M T}\left(\|v_0\|^2 + C_0\right). \tag{2.35}$$

We recall that $\alpha_m = \min_{x\in\Gamma} \alpha(x)$ was defined in (2.11) and is positive.

Proposition 2.5. *Let us assume (2.31) and let $y_0 \in L^2(\Omega)$, $y_0(x) \leq y_s$ a.e. $x \in \Omega$. Then (2.23) has a unique solution (y_i^h, ζ_i^h), $\zeta_i^h \in \beta^*(y_i^h)$ a.e. on Q, $y_i^h \in D(A^h)$ and the discretization scheme is stable, i.e.,*

$$\left\| y_p^h \right\|^2 \leq C, \ for \ any \ p = 1, \ldots, n, \tag{2.36}$$

$$h \sum_{i=1}^{p} \left\| \zeta_i^h \right\|_V^2 \leq C(\alpha_m), \ for \ any \ p = 1, \ldots, n, \tag{2.37}$$

$$h \sum_{i=1}^{p} \left\| \frac{y_i^h - y_{i-1}^h}{h} \right\|_{V'}^2 \leq C(\alpha_m), \ for \ any \ p = 1, \ldots, n, \tag{2.38}$$

where by C and $C(\alpha_m)$ we have denoted several constants depending on the problem data and $\alpha_m^{-1/2}$, respectively and independent on p and h.

Proof. By (2.22) and (2.10) we have $f_i^h \in L^2(\Omega)$, and

$$h \sum_{i=1}^{p} \left\| f_i^h \right\|^2 \leq \sum_{i=1}^{n} \int_{(i-1)h}^{ih} \left\| f(s) \right\|^2 ds = \int_0^T \left\| f(s) \right\|^2 ds := C^f.$$

Since for h satisfying (2.31) the operator $\frac{1}{h}I + A^h$ is invertible and has a Lipschitz continuous inverse it follows that (2.23) has a unique solution $y_i^h \in D(A^h)$, meaning in fact that $y_i^h \in L^2(\Omega)$, $\zeta_i^h \in V$, $\zeta_i^h \in \beta^*(y_i^h)$ for all $i = 1, \ldots, n$. Next, we shall establish the estimates which will ensure the scheme stability.

We write (2.25) for $\psi = y_i^h$. By (2.32) and (2.9) we have that

$$\frac{\left\| y_i^h \right\|^2}{2h} - \frac{\left\| y_{i-1}^h \right\|^2}{2h} + \sqrt{h} \left\| y_i^h \right\|_V^2 \leq \langle f_i^h, y_i^h \rangle_{V',V} + \int_\Omega a(x) y_i^h \cdot \nabla y_i^h dx$$

$$\leq \int_\Omega f_i^h y_i^h dx + \frac{1}{2} \int_\Omega a(x) \cdot \nabla \left(y_i^h \right)^2 dx$$

$$\leq \left\| f_i^h \right\| \left\| y_i^h \right\| + \frac{1}{2} \int_\Omega \nabla \cdot (a(x)(y_i^h)^2) dx - \frac{1}{2} \int_\Omega (y_i^h)^2 \nabla \cdot a(x) dx$$

$$\leq \frac{1}{2} \left\| f_i^h \right\|^2 + \frac{1}{2} \left\| y_i^h \right\|^2 + \frac{1}{2} \left\| a \right\|_{1,\infty} \left\| y_i^h \right\|^2,$$

where $\left\| a \right\|_{1,\infty} = \sum_{i=1}^{n} \left\| a_i \right\|_{W^{1,\infty}(\Omega)}$. By summing up as $i = 1, \ldots, p$ we obtain

$$\|y_p^h\|^2 + \sqrt{h}h\sum_{i=1}^{p}\|y_i^h\|_V^2 \le h\sum_{i=1}^{p}\|f_i^h\|^2 + \|y_0^h\|^2 + (\|a\|_{1,\infty} + 1)h\sum_{i=1}^{p}\|y_i^h\|^2$$

$$(2.39)$$

and using Lemma 2.4 (i.e., (2.34) and (2.35)) with

$$C_M = \|a\|_{1,\infty} + 1, \ C_0 = \int_0^T \|f(t)\|^2\, dt + \|y_0\|^2 \tag{2.40}$$

we get both (2.36) and

$$h\sum_{i=1}^{p}\|y_i^h\|^2 \le C(\|y_0\|^2 + \int_0^T \|f(t)\|^2\, dt). \tag{2.41}$$

This plugged into (2.39) leads to

$$\sqrt{h}h\sum_{i=1}^{p}\|y_i^h\|_V^2 \le C \text{ for any } p = 1,\ldots,n. \tag{2.42}$$

By C we have denoted several positive constants and we retain another estimate for the time step

$$h < \frac{1}{\|a\|_{1,\infty} + 1} \tag{2.43}$$

implied by the condition $h < \frac{1}{C_M}$, occurring in the proof of Lemma 2.4.

We prove now (2.37). Since (y_i^h, ζ_i^h) is the solution we write (2.25) for $\psi = \zeta_i^h$, where $\zeta_i^h(x) \in \beta^*(y_i^h(x))$ a.e. $x \in \Omega$,

$$\frac{1}{h}\int_\Omega (y_i^h - y_{i-1}^h)\,\zeta_i^h dx + \int_\Omega |\nabla\zeta_i^h|^2\, dx + \int_\Gamma \alpha(x)\left(\zeta_i^h\right)^2 d\sigma$$

$$+\sqrt{h}\int_\Omega \nabla y_i^h \cdot \nabla\zeta_i^h dx + \sqrt{h}\int_\Gamma \alpha(x)y_i^h \zeta_i^h d\sigma \tag{2.44}$$

$$= \int_\Omega f_i^h\zeta_i^h dx + \int_\Omega a(x)y_i^h \cdot \nabla\zeta_i^h dx \le 2kc_H^2 \|f_i^h\|^2 + \frac{1}{k}\|\zeta_i^h\|_V^2$$

$$+2a_0^2 k \|y_i^h\|^2$$

where we have used the relation (2.14). But $\partial j = \beta^*$, so we can write

$$\int_\Omega (j(y_i^h) - j(y_{i-1}^h))dx \le \int_\Omega \zeta_i^h(y_i^h - y_{i-1}^h)dx. \tag{2.45}$$

Summing up (2.44) from $i = 1$ to p we obtain

$$\int_\Omega j(y_p^h)dx + \left(1 - \frac{1}{k}\right) h \sum_{i=1}^p \left\|\zeta_i^h\right\|_V^2 \tag{2.46}$$

$$\leq 2a_0^2 kh \sum_{i=1}^p \left\|y_i^h\right\|^2 + 2kc_H^2 \int_0^T \|f(t)\|^2 \, dt + \int_\Omega j(y_0)dx.$$

We recall that $y_0^h = y_0$ with $y_0 \leq y_s$ and $\beta^*(r) < \beta_s^*$ for $r < y_s$ and so

$$\int_\Omega j(y_0)dx \leq \lim_{y \nearrow y_s} \int_\Omega \int_0^y \beta^*(\sigma)d\sigma dx \leq \beta_s^* y_s \mathrm{meas}(\Omega) < \infty. \tag{2.47}$$

Taking also into account (2.41) the right-hand side in (2.46) turns out to be bounded and we obtain for any $p = 1, \ldots, n$, that

$$\int_\Omega j(y_p^h)dx + \left(1 - \frac{1}{k}\right) h \sum_{i=1}^p \left\|\zeta_i^h\right\|_V^2 \leq C(\alpha_m) \left(\|y_0\|^2 + \int_0^T \|f(t)\|^2 \, dt + 1\right), \tag{2.48}$$

(where $C(\alpha_m)$ is another constant depending on $\beta_s^*, y_s, k, T, \|a\|_{1,\infty}, \alpha_m^{-1/2}$). We also add that in the degenerate case hypothesis (2.4) implies that

$$j(r) \geq \frac{\gamma_\beta}{(m+1)(m+2)} |r|^{m+2} \geq 0 \text{ for } r \leq 0 \tag{2.49}$$

and (2.48) leads to (2.37), and

$$\int_\Omega j(y_p^h)dx \leq C(\alpha_m) \left(\|y_0\|^2 + \int_0^T \|f(t)\|^2 \, dt + 1\right). \tag{2.50}$$

From here we deduce that $y_p^h \leq y_s$ a.e. $x \in \Omega$ for any $p = 1, \ldots, n$.

Finally we pass to show (2.38). We multiply (2.23) by $\delta y_i^h := \frac{y_i^h - y_{i-1}^h}{h}$ scalarly in V'. Using again (2.45) we get

$$\left\|\delta y_i^h\right\|_{V'}^2 + \frac{1}{h} \int_\Omega \left(j(y_i^h) - j(y_{i-1}^h)\right) dx + \sqrt{h} \int_\Omega y_i^h \delta y_i^h dx$$

$$\leq \left\|f_i^h\right\|_{V'} \left\|\delta y_i^h\right\|_{V'} + a_0 \left\|y_i^h\right\| \left\|\delta y_i^h\right\|_{V'}$$

$$\leq \frac{1}{k} \left\|\delta y_i^h\right\|_{V'}^2 + 3kc_H^2 \left\|f_i^h\right\|^2 + 3M_K^2 k \left\|y_i^h\right\|^2 + 3kh \left\|y_i^h\right\|_V^2.$$

Next we have

$$\|\delta y_i^h\|_{V'}^2 + \frac{1}{h} \int_\Omega \left(j(y_i^h) - j(y_{i-1}^h) \right) dx$$

$$\leq \|f_i^h\|_{V'} \|\delta y_i^h\|_{V'} + a_0 \|y_i^h\| \|\delta y_i^h\|_{V'} + \sqrt{h} \left| \langle \delta y_i^h, y_i^h \rangle_{V',V} \right|$$

$$\leq \frac{1}{k} \|\delta y_i^h\|_{V'}^2 + 3kc_H^2 \|f_i^h\|^2 + 3a_0^2 k \|y_i^h\|^2 + 3kh \|y_i^h\|_V^2,$$

and so we obtain

$$\int_\Omega \left(j(y_i^h) - j(y_{i-1}^h) \right) dx + \left(1 - \frac{1}{k} \right) h \|\delta y_i^h\|_{V'}^2$$

$$\leq 3kc_H^2 h \|f_i^h\|^2 + 3ka_0^2 h \|y_i^h\|^2 + 3kh \cdot h \|y_i^h\|_V^2.$$

Summing up with respect to $i = 1, \ldots, n$ and taking into account (2.41) and (2.42) we get that

$$\int_\Omega j(y_p^h) dx + \left(1 - \frac{1}{k} \right) h \sum_{i=1}^p \|\delta y_i^h\|_{V'}^2 \leq C(\alpha_m). \tag{2.51}$$

Thus (2.38) is proved and the proof of Proposition 2.5 is ended.   □

We underline that by C we denote several constants depending on the problem data, while $C(\alpha_m)$ depends in addition, via (2.14), on $\alpha_m^{-1/2}$, with $\alpha_m > 0$ (see (2.11)). That is why we cannot deduce from here the estimates for Neumann boundary conditions (corresponding to $\alpha \equiv 0$). For that we must follow another way (see [89]).

2.1.3   Convergence of the Discretization Scheme

We have now all results required to prove the scheme convergence. We define the piecewise constant functions

$$\begin{aligned} y^h(t, x) &= y_i^h(x), && \text{for } t \in ((i-1)h, ih], \\ \zeta^h(t, x) &= \zeta_i^h(x), && \text{for } t \in ((i-1)h, ih], \\ f^h(t, x) &= f_i^h(x), && \text{for } t \in ((i-1)h, ih], \end{aligned} \tag{2.52}$$

for $i = 1, \ldots, n$, and

$$y^h(t, x) = y_0(x), \quad \text{for } t \in [-h, 0].$$

Using (2.36)–(2.38), (2.48) and (2.42) we deduce the following estimates:

$$\|y^h(t)\| \le C \text{ for any } t \in [0, T], \tag{2.53}$$

$$\int_0^T \|\zeta^h(t)\|_V^2\, dt \le C(\alpha_m), \ \zeta^h(t, x) \in \beta^*(y^h(t, x)) \text{ a.e. } (t, x) \in Q, \tag{2.54}$$

$$\int_0^T \left\|\frac{y^h(t) - y^h(t-h)}{h}\right\|_{V'}^2\, dt \le C(\alpha_m), \tag{2.55}$$

$$\int_\Omega j(y^h(t))dx \le C(\alpha_m), \text{ for } t \in [0, T], \tag{2.56}$$

$$h^{\frac{1}{2}} \int_0^T \|y^h(t)\|_V^2\, dt \le C. \tag{2.57}$$

Recalling that $y^h(0) = y_0^h = y_0$, by Proposition 2.5 we have proved in fact that y^h is a h-approximate solution to the Cauchy problem (2.20)–(2.21). Also we recall that $\alpha_m > 0$, defined by (2.11).

Theorem 2.6. *Assume (2.31) and let $y_0 \in L^2(\Omega)$, $y_0(x) \le y_s$ a.e. $x \in \Omega$. Then problem (2.20)–(2.21) has at least a mild solution obtained as the weak limit of the sequence $(y^h)_{h>0}$,*

$$y = \lim_{h \to 0} y^h \text{ weakly in } L^2(0, T; L^2(\Omega)). \tag{2.58}$$

Moreover,

$$y \in L^2(0, T; L^2(\Omega)) \cap W^{1,2}([0, T]; V'), \tag{2.59}$$

$$y(t, x) \le y_s \text{ a.e. } (t, x) \in Q,$$

and it satisfies the estimates

$$\|y(t)\| \le C \text{ for any } t \in [0, T], \tag{2.60}$$

$$\int_0^T \|\zeta(t)\|_V^2\, dt \le C(\alpha_m), \ \zeta(t, x) \in \beta^*(y(t, x)) \text{ a.e. } (t, x) \in Q, \tag{2.61}$$

$$\int_0^T \left\|\frac{dy}{dt}(t)\right\|_{V'}^2\, dt \le C(\alpha_m), \tag{2.62}$$

$$\int_\Omega j(y(t))dx \le C(\alpha_m), \text{ for any } t \in [0, T]. \tag{2.63}$$

Proof. By (2.53)–(2.57) it follows that there exist y, ζ, κ, χ in appropriate spaces and we can select a subsequence of $(y^h)_{h>0}$ (denoted in the same way), such that

$$y^h \overset{w*}{\to} y \text{ in } L^\infty(0,T;L^2(\Omega)) \text{ as } h \to 0, \tag{2.64}$$

$$\zeta^h \rightharpoonup \zeta \text{ in } L^2(0,T;V) \text{ as } h \to 0, \tag{2.65}$$

$$h^{\frac{1}{4}} y^h \rightharpoonup \kappa \text{ in } L^2(0,T;V) \text{ as } h \to 0, \tag{2.66}$$

$$\frac{y^h(t) - y^h(t-h)}{h} \rightharpoonup \chi \text{ in } L^2(0,T;V') \text{ as } h \to 0. \tag{2.67}$$

We show next that $\chi = \frac{dy}{dt}$ in the sense of distributions. To this end we take $\varphi \in C_0^\infty(Q)$ and compute

$$\int_0^T \left\langle \frac{y^h(t) - y^h(t-h)}{h}, \varphi(t) \right\rangle_{V',V} dt$$

$$= \int_0^T \frac{1}{h} \langle y^h(t), \varphi(t) \rangle_{V',V} dt - \int_0^T \frac{1}{h} \langle y^h(t-h), \varphi(t) \rangle_{V',V} dt$$

$$= \int_0^{T-h} \frac{1}{h} \langle y^h(t), \varphi(t) \rangle_{V',V} dt + \int_{T-h}^T \frac{1}{h} \langle y^h(t), \varphi(t) \rangle_{V',V} dt$$

$$- \int_0^h \frac{1}{h} \langle y^h(t-h), \varphi(t) \rangle_{V',V} dt - \int_h^T \frac{1}{h} \langle y^h(t-h), \varphi(t) \rangle_{V',V} dt.$$

We make the function transformation $t \to t-h$ in the last integral and get

$$\int_0^T \left\langle \frac{y^h(t) - y^h(t-h)}{h}, \varphi(t) \right\rangle_{V',V} dt$$

$$= \int_0^{T-h} \frac{1}{h} \langle y^h(t), \varphi(t) \rangle_{V',V} dt + \int_{T-h}^T \frac{1}{h} \langle y^h(t), \varphi(t) \rangle_{V',V} dt$$

$$- \int_0^h \frac{1}{h} \langle y^h(t-h), \varphi(t) \rangle_{V',V} dt - \int_0^{T-h} \frac{1}{h} \langle y^h(t), \varphi(t+h) \rangle_{V',V} dt$$

$$= - \int_0^{T-h} \left\langle y^h(t), \frac{\varphi(t+h) - \varphi(t)}{h} \right\rangle_{V',V} dt + \int_{T-h}^T \frac{1}{h} \langle y^h(t), \varphi(t) \rangle_{V',V} dt$$

$$- \int_0^h \frac{1}{h} \langle y^h(t-h), \varphi(t) \rangle_{V',V} dt.$$

Passing to the limit as $h \to 0$ and taking into account that the two last integrals vanish due to the fact that φ has the support compact in Q, we obtain

$$\lim_{h \to 0} \int_0^T \left\langle \frac{y^h(t) - y^h(t-h)}{h}, \varphi(t) \right\rangle_{V',V} dt = - \int_0^T \left\langle y(t), \frac{d\varphi}{dt}(t) \right\rangle_{V',V} dt,$$

which proves that $\dfrac{y^h(t) - y^h(t-h)}{h} \rightharpoonup \chi = \frac{dy}{dt}$ in the sense of distributions as $h \to 0$.

From (2.56) and the lower semicontinuity property of the convex function j we also get that

$$\int_\Omega j(y(t))dx \leq \lim_{h \to 0} \inf \int_\Omega j(y^h(t))dx \leq C(\alpha_m) \qquad (2.68)$$

which implies that $y(t,x) \leq y_s$ a.e. on Q. By (2.66) and the trace continuity it follows the trace convergence

$$h^{\frac{1}{4}} y^h \Big|_\Sigma \rightharpoonup \kappa|_\Sigma \text{ in } L^2(0,T;L^2(\Gamma)) \text{ as } h \to 0$$

and so

$$h^{\frac{1}{2}} y^h \Big|_\Sigma \to 0 \text{ in } L^2(0,T;L^2(\Gamma)) \text{ as } h \to 0. \qquad (2.69)$$

The next step is to prove that $y^h(t) \to y(t)$ in V' as $h \to 0$.

We denote by $BV([0,T];V')$ the set of functions with bounded variation from $[0,T]$ to V' and show that $y^h \in BV([0,T];V')$, i.e.,

$$V_0^T(y^h) = \sup_{P \in \mathcal{P}} \sum_{i=1}^{n_p} \|y^h(s_i) - y^h(s_{i-1})\|_{V'} \leq C(\alpha_m), \qquad (2.70)$$

where $\mathcal{P} = \{P(s_i) = (s_0, \ldots, s_{n_p}); P \text{ is a partition of } [0,T]\}$ is the set of all partitions of $[0,T]$.

Here is the argument. First, let us take an equidistant partition, considering for example that $s_i = t_i$ (i.e., the partition considered up to now). We have that

$$\left(\sum_{i=1}^n \|y^h(t_i) - y^h(t_{i-1})\|_{V'} \right)^2 \qquad (2.71)$$

$$\leq n \sum_{i=1}^n \|y_i^h - y_{i-1}^h\|_{V'}^2 = nh \cdot h \sum_{i=1}^n \left\| \frac{y_i^h - y_{i-1}^h}{h} \right\|_{V'}^2 \leq TC(\alpha_m),$$

by (2.38).

If the partition is not equidistant, having for some i, $s_{i-1} \in (t_{i-1}, t_i)$ and $s_i \in (t_{p-1}, t_p)$, with $p > i$ (and $s_l = t_l$ for $l \neq i$), then we can write

$$\left\|y^h(s_i) - y^h(s_{i-1})\right\|_{V'} \le \left\|y^h(t_i) - y^h(s_{i-1})\right\|_{V'}$$

$$+ \sum_{l=i+1}^{p-1} \left\|y^h(t_l) - y^h(t_{l-1})\right\|_{V'} + \left\|y^h(s_i) - y^h(t_{p-1})\right\|_{V'},$$

where the first and the last terms on the right-hand side vanish. Hence

$$\left\|y^h(s_i) - y^h(s_{i-1})\right\|_{V'} \le \sum_{l=i+1}^{p-1} \left\|y^h(t_l) - y^h(t_{l-1})\right\|_{V'}$$

$$+ \left\|y^h(t_i) - y^h(t_{i-1})\right\|_{V'} + \left\|y^h(t_p) - y^h(t_{p-1})\right\|_{V'},$$

so that (2.70) follows by changing the partition of $[0,T]$, by introducing between s_{i-1} and s_i the points $t_i, \ldots, t_{p-1}$ and applying again (2.71) for the new partition.

If $s_i, s_{i-1} \in (t_{i-1}, t_i)$ for some i, then $\left\|y^h(s_i) - y^h(s_{i-1})\right\|_{V'} = 0$.

Since we have (2.70), (2.53) and $L^2(\Omega)$ is compact in V' we can apply the infinite dimensional Helly theorem (see [21], Theorem 3.5 and Remark 3.2, pp. 60) to obtain that $y \in BV([0,T]; V')$ and

$$y^h(t) \to y(t) \text{ in } V' \text{ for all } t \in [0,T]. \tag{2.72}$$

From here and (2.64) we deduce that for $a_i \in W^{1,\infty}(\Omega)$ we have

$$a_i y^h(t) \to a_i y(t) \text{ in } V' \text{ for } t \in [0,T], \tag{2.73}$$

$$a_i y^h \rightharpoonup a_i y \text{ in } L^2(0,T; L^2(\Omega)) \text{ as } h \to 0. \tag{2.74}$$

At the end we assert that

$$f^h \to f \text{ in } C([0,T]; L^2(\Omega)). \tag{2.75}$$

Indeed, let $f \in C([0,T]; L^2(\Omega))$ and $s,t \in (t_{i-1}, t_i)$. Then, since $|s - t| \le h$, it follows that $\|f(s) - f(t)\| \le \delta(h)$ where $\delta(h) \to 0$, as $h \to 0$, whence

$$\left\|f^h(t) - f(t)\right\| = \left\|\frac{1}{h}\int_{t_{i-1}}^{t_i} (f(s) - f(t))ds\right\| \le \frac{1}{h}\int_{t_{i-1}}^{t_i} \|f(s) - f(t)\| \, ds \le \delta(h),$$

for any $t \in [0,T]$, which implies (2.75).

Now, writing (2.25) for $i = 1, \ldots, n$, and summing up with respect to i we get

$$\int_0^T \left\langle \frac{y^h(t) - y^h(t-h)}{h}, \psi(t) \right\rangle_{V',V} dt + h^{\frac{1}{4}} \int_Q h^{\frac{1}{4}} \nabla y^h \cdot \nabla \psi dx dt$$

$$+ h^{\frac{1}{4}} \int_\Sigma \alpha(x) h^{\frac{1}{4}} y^h \psi d\sigma dt + \int_Q (\nabla \zeta^h - a(x)y^h) \cdot \nabla \psi dx dt$$

$$+ \int_\Sigma \alpha(x) \zeta^h \psi d\sigma dt = \int_0^T \int_\Omega f^h \psi dx dt, \text{ for any } \psi \in L^2(0, T; V).$$

Then we pass to the limit as $h \to 0$, using (2.67), (2.65), (2.66), (2.69), (2.74) and (2.75) and deduce the equation

$$\int_0^T \left\langle \frac{dy}{dt}(t), \psi(t) \right\rangle_{V',V} dt + \int_Q (\nabla \zeta - a(x)y) \cdot \nabla \psi dx dt \qquad (2.76)$$

$$+ \int_\Sigma \alpha(x) \zeta \psi d\sigma dt = \int_Q f \psi dx dt \text{ for any } \psi \in L^2(0, T; V).$$

We stress that (2.66) implied $h^{\frac{1}{2}} y^h \rightharpoonup 0$ in $L^2(0, T; V)$ as $h \to 0$. If we take $\phi \in C_0^\infty(Q)$ in the previous equation we obtain that (y, ζ) satisfy

$$\frac{dy}{dt} - \Delta\zeta + \nabla \cdot (a(x)y) = f \text{ in } \mathcal{D}'(Q). \qquad (2.77)$$

It remains to show that $\zeta \in \beta^*(y)$ a.e. on Q and to this end we prove that

$$\limsup_{h \to 0} \int_0^T (\zeta^h(t), y^h(t)) dt \leq \int_0^T (\zeta(t), y(t)) dt. \qquad (2.78)$$

We multiply (2.23) scalarly in V' by y_i^h and obtain

$$\frac{\|y_i^h\|_{V'}^2}{2h} - \frac{\|y_{i-1}^h\|_{V'}^2}{2h} + \int_\Omega \zeta_i^h y_i^h dx + \sqrt{h} \int_\Omega (y_i^h)^2 dx \qquad (2.79)$$

$$\leq \int_\Omega a(x) y_i^h \cdot \nabla \psi_i^h dx + \int_\Omega f_i^h y_i^h dx,$$

where $\psi_i^h \in V$ is the solution to $A_\Delta \psi_i^h = y_i^h$, i.e.,

$$- \Delta \psi_i^h = y_i^h, \quad \frac{\partial \psi_i^h}{\partial \nu} + \alpha \psi_i^h = 0 \text{ on } \Gamma \text{ for each } i. \qquad (2.80)$$

We denote $\psi^h(t) = \psi_i^h$ for $t \in (t_{i-1}, t_i]$ and then we can rewrite this problem as

$$- \Delta \psi^h(t) = y^h(t), \quad \frac{\partial \psi^h}{\partial \nu}(t) + \alpha \psi^h(t) = 0 \text{ on } \Gamma \text{ a.e. } t \in (0, T), \quad (2.81)$$

with $y^h(t) \in L^2(\Omega)$.

By the elliptic regularity (see [1]) this problem has a unique solution $\psi^h(t) \in H^2(\Omega)$, a.e. $t \in (0, T)$, with $\left\| \psi^h(t) \right\|_{H^2(\Omega)} \leq c_0 \left\| y^h(t) \right\| \leq C$, due to (2.36), and since $H^2(\Omega)$ is compact in V we get that

$$\psi^h(t) \to \psi(t) \text{ in } V, \text{ a.e. } t \in (0, T). \quad (2.82)$$

Hence passing to the limit in (2.81) we get that the function ψ satisfies the problem

$$A_\Delta \psi(t) = y(t) \text{ for } t \in [0, T] \quad (2.83)$$

with A_Δ defined in (2.16). Recalling that $\frac{dy}{dt} \in L^2(0, T; V')$ we obtain

$$A_\Delta \frac{d\psi}{dt}(t) = \frac{dy}{dt}(t) \text{ a.e. } t \in (0, T) \quad (2.84)$$

and so it follows that $\psi \in W^{1,2}([0, T]; V)$. From here and (2.83) we get

$$\left\langle \frac{dy}{dt}(t), \psi(t) \right\rangle_{V', V} = \int_\Omega \nabla \frac{d\psi}{dt}(t) \cdot \nabla \psi(t) dx + \int_\Gamma \alpha \frac{d\psi}{dt}(t) \psi(t) d\sigma$$

$$= \frac{1}{2} \frac{d}{dt} \left\| \psi(t) \right\|_V^2 = \frac{1}{2} \frac{d}{dt} \left\| y(t) \right\|_{V'}^2.$$

Finally, we deduce that

$$\int_0^T \left\langle \frac{dy}{dt}(t), \psi(t) \right\rangle_{V', V} dt = \frac{1}{2} \left(\left\| y(T) \right\|_{V'}^2 - \left\| y(0) \right\|_{V'}^2 \right). \quad (2.85)$$

Now, we sum up (2.79) with respect to $i = 1, \ldots, n$ and get

$$\frac{\left\| y_n^h \right\|_{V'}^2}{2h} - \frac{\left\| y_0^h \right\|_{V'}^2}{2h} + \sum_{i=1}^n \int_\Omega \zeta_i^h y_i^h dx \leq \sum_{i=1}^n \int_\Omega a(x) y_i^h \cdot \nabla \psi_i^h dx + \sum_{i=1}^n \int_\Omega f_i^h y_i^h dx$$

whence we obtain

$$h \sum_{i=1}^n \int_\Omega \zeta_i^h y_i^h dx \leq -\frac{\left\| y_n^h \right\|_{V'}^2}{2} + \frac{\left\| y_0^h \right\|_{V'}^2}{2} + h \sum_{i=1}^n \int_\Omega a(x) y_i^h \cdot \nabla \psi_i^h dx$$

$$+ h \sum_{i=1}^n \int_\Omega f_i^h y_i^h dx.$$

We still can write

$$\int_0^T (\zeta^h(t), y^h(t)) \, dt \leq -\frac{\|y^h(T)\|_{V'}^2}{2} + \frac{\|y^h(0)\|_{V'}^2}{2} \tag{2.86}$$

$$+ \int_0^T \int_\Omega a(x) y^h(t) \cdot \nabla \psi^h(t) dx dt + \int_0^T \int_\Omega f^h(t) y^h(t) dx dt.$$

By (2.75) and (2.64) we get

$$\int_0^T \int_\Omega f^h(t) y^h(t) dt \to \int_0^T \int_\Omega f(t) y(t) dt \text{ as } h \to 0.$$

We notice that

$$\int_0^T \int_\Omega a(x) y^h(t) \cdot \nabla \psi^h(t) dt \to \int_0^T \int_\Omega a(x) y(t) \cdot \nabla \psi(t) dt \text{ as } h \to 0$$

because of (2.64) and (2.82). Therefore, by passing to the limit in (2.86) and using all previous convergencies and $y^h(0) = y_0$ we get

$$\limsup_{h \to 0} \int_0^T (\zeta^h(t), y^h(t)) \, dt \leq -\frac{\|y(T)\|_{V'}^2}{2} + \frac{\|y(0)\|_{V'}^2}{2}$$

$$+ \int_0^T \int_\Omega a(x) y(t) \cdot \nabla \psi(t) dx dt + \int_0^T \int_\Omega f(t) y(t) dx dt$$

$$= - \int_0^T \left\langle \frac{dy}{dt}(t), \psi(t) \right\rangle_{V',V} dt + \int_0^T \int_\Omega a(x) y(t) \cdot \nabla \psi(t) dx dt$$

$$+ \int_0^T \int_\Omega f y \, dx dt = \int_0^T (\zeta(t), y(t)) \, dt.$$

Thus, (2.78) is proved. Hence, we have $y^h \rightharpoonup y$, $\zeta^h \rightharpoonup \zeta$ in $L^2(0, T; L^2(\Omega))$ as $h \to 0$, $\zeta^h \in \beta^*(y^h)$ and β^* is maximal monotone. In conclusion, according to a result given in [14], pp. 41, we deduce that $\zeta \in \beta^*(y)$ a.e. on Q. Since $\zeta \in \beta^*(y) \in L^2(0, T; V)$, it follows that $y \leq y_s$ a.e. on Q.

Finally, estimates (2.60)–(2.63) are obtained by passing to the limit as $h \to 0$ in (2.53)–(2.56) on the basis of weakly lower semicontinuity (of the norms and j) and the weakly convergencies (2.64), (2.65), (2.67).

Comparing (2.76) where $\zeta \in \beta^*(y)$ a.e. on Q with (2.18) we notice that the mild solution to (2.20)–(2.21) obtained by this proof is in fact a weak solution to (2.1)–(2.3). □

We notice here the necessity of the linear dependence of K_0 on the solution. In the case of a nonlinear dependence $K_0(x, y) = a(x) K(y)$ as in Chap. 1, the weak convergence $y^h \rightharpoonup y$ only and the Lipschitz property of $K(y)$ are not sufficient to get more information about the weak limit of $(K(y^h))_{h>0}$, i.e., to show that it is $K_0(x, y)$.

2.1.4   Uniqueness

We shall give some results for the solution uniqueness in particular cases.

Proposition 2.7. *Under the hypotheses of Theorem 2.6 with $N = 1$ and*

$$a \cdot \nu = 0 \text{ on } \Gamma, \tag{2.87}$$

the solution to problem (2.20)–(2.21) is unique.

If $a = 0$ the solution to the N-dimensional problem (2.20)–(2.21) is unique.

Proof. We consider two problems (2.20)–(2.21) with the data $\{y_0, f\}$ and $\{\overline{y_0}, \overline{f}\}$ having the solutions y and $\overline{y}$, and multiply the difference of the corresponding equations by $y - \overline{y}$ scalarly in V'. We have

$$\frac{1}{2} \|y(t) - \overline{y}(t)\|_{V'}^2 - \frac{1}{2} \|y_0 - \overline{y}_0\|_{V'}^2 \tag{2.88}$$

$$\leq \int_0^t \int_\Omega \left\{ a(x)(y(\tau) - \overline{y}(\tau)) \cdot \nabla \psi(\tau) + (f(\tau) - \overline{f}(\tau))(y(\tau) - \overline{y}(\tau)) \right\} dx d\tau$$

where $\psi(t)$ is the solution to $A_\Delta \psi(t) = y(t) - \overline{y}(t)$, for $t \in [0, T]$.

In the case $N = 1$ we can compute the term $\int_\Omega a(x) v \phi_x dx$, where $A_\Delta \phi = v$ for any $v \in L^2(\Omega)$. We recall that the latter equation is equivalent with

$$-\phi_{xx} = v \text{ in } \Omega, \quad \frac{\partial \phi}{\partial \nu} + \alpha \phi = 0 \text{ on } \Gamma.$$

We multiply the first equation in this problem by $a(x)\phi_x$ scalarly in $L^2(\Omega)$

$$\int_\Omega a(x) v \phi_x dx = -\int_\Omega a(x) \phi_x \phi_{xx} dx = -\frac{1}{2} \int_\Gamma a(x) \cdot \nu \phi_x^2 d\sigma + \frac{1}{2} \int_\Omega a_x \phi_x^2 dx.$$

Using (2.87) we obtain

$$\int_\Omega a(x) v \phi_x dx = \frac{1}{2} \int_\Omega a_x \phi_x^2 dx \leq \frac{1}{2} \|a\|_{1,\infty} \|v\|_{V'}^2 \tag{2.89}$$

which applied for $v = y(\tau) - \overline{y}(\tau)$ gives

$$\int_\Omega a(x)(y(\tau) - \overline{y}(\tau)) \cdot \nabla \psi(\tau) dx \leq \frac{1}{2} \|a\|_{1,\infty} \|y(\tau) - \overline{y}(\tau)\|_{V'}^2. \tag{2.90}$$

Returning to (2.88) and considering the same data $(y_0 = \overline{y}_0,\ f = \overline{f})$ we obtain that

$$\|y(t) - \overline{y}(t)\|_{V'}^2 \le \|a\|_{1,\infty} \int_0^t \|y(\tau) - \overline{y}(\tau)\|_{V'}^2\, d\tau. \tag{2.91}$$

Applying Gronwall's lemma we get that the solution is unique.

In the N-dimensional case, with $a = 0$ the uniqueness follows immediately from (2.88) for the same data.   □

In the general N-dimensional case, in the absence of a uniqueness result, the convergence of y^h to y takes place on a subsequence of $(y^h)_{h>0}$.

2.1.5   Error Estimate

Proposition 2.8. *Let $N = 1$ and assume the hypotheses of Theorem 2.6 and (2.87). Then*

$$\left\|y(t_i) - y_i^h\right\|_{V'} = O(h^{1/4}) \ as \ h \to 0, \ for \ i = 1,\dots,n. \tag{2.92}$$

Proof. By Theorem 2.6 and (2.87) problem (2.20)–(2.21) has a unique solution. Since (2.20) takes place in V', a.e. t, we can integrate it with respect to t on (t_{i-1}, t_i), according to the integration of vectorial functions. More exactly, if $g \in L^2(0, T; V')$ we define

$$\left\langle \int_{t_1}^{t_2} g(t)dt, \psi \right\rangle_{V',V} := \int_{t_1}^{t_2} \langle g(t), \psi \rangle_{V',V}\, dt, \ \text{for any } \psi \in V$$

and see that $\int_{t_1}^{t_2} g(t)dt$ is well defined as an element of V'. By integrating (2.20) on (t_{i-1}, t_i) and dividing by h we get the equation in V'

$$\frac{y(t_i) - y(t_{i-1})}{h} + \frac{1}{h}\int_{t_{i-1}}^{t_i} Ay(t)dt = \frac{1}{h}\int_{t_{i-1}}^{t_i} f(t)dt.$$

We subtract the corresponding equations from the discretized system (2.23) and denote $w_i = y(t_i) - y_i^h$. Recalling (2.22) we get

$$\frac{w_i - w_{i-1}}{h} - \frac{1}{h}\int_{t_{i-1}}^{t_i} (Ay(t) - A^h y_i^h)dt = 0. \tag{2.93}$$

We multiply (2.93) scalarly in V' by w_i and obtain after some computations

$$\frac{\|w_i\|_{V'}^2}{h} + \frac{1}{h}\int_{t_{i-1}}^{t_i}\int_\Omega (\zeta(t) - \zeta_i^h)w_i dx dt - \frac{1}{h}\int_{t_{i-1}}^{t_i}\int_\Omega \sqrt{h}y_i^h w_i dx dt$$

$$= \frac{1}{h}\int_{t_{i-1}}^{t_i}\int_\Omega a(x)(y(t) - y_i^h)\cdot\nabla\psi dx dt + \frac{1}{h}\|w_i\|_{V'}\|w_{i-1}\|_{V'},$$

where ψ satisfies $A_\Delta\psi = w_i$.

As specified before, because $w_i \in L^2(\Omega)$, this problem has a unique solution $\psi \in H^2(\Omega)$, with $\|\psi\|_V \le \|w_i\|_{V'}$ and $\|\psi_{xx}\| \le c\|w_i\|$. Next we can write

$$\frac{\|w_i\|_{V'}^2}{2h} - \frac{\|w_{i-1}\|_{V'}^2}{2h} + \frac{1}{h}\int_{t_{i-1}}^{t_i}\int_\Omega (\zeta(t) - \zeta_i^h)(y(t) - y_i^h)dx dt$$

$$+\frac{1}{h}\int_{t_{i-1}}^{t_i}\int_\Omega (\zeta(t) - \zeta_i^h)(y(t_i) - y(t))dx dt$$

$$\le \frac{1}{h}\int_{t_{i-1}}^{t_i}\langle a(x)(y(t) - y(t_i)), \nabla\psi\rangle_{V',V} dt$$

$$+\frac{1}{h}\int_{t_{i-1}}^{t_i}\langle a(x)(y(t_i) - y_i^h), \nabla\psi\rangle_{V',V} dt + \frac{1}{h}\int_{t_{i-1}}^{t_i}\left\langle w_i, \sqrt{h}y_i^h\right\rangle_{V',V} dt.$$

Using (2.89) with $v = y(t_i) - y_i^h = w_i$ we get

$$\frac{\|w_i\|_{V'}^2}{2h} - \frac{\|w_{i-1}\|_{V'}^2}{2h} + \frac{1}{h}\int_{t_{i-1}}^{t_i}\int_\Omega (\zeta(t) - \zeta_i^h)(y(t) - y_i^h)dx dt$$

$$\le \frac{1}{h}\int_{t_{i-1}}^{t_i}\|y(t_i) - y(t)\|_{V'}\|\zeta(t) - \zeta_i^h\|_V dt \qquad (2.94)$$

$$+\frac{1}{h}\int_{t_{i-1}}^{t_i}\|a(x)(y(t) - y(t_i))\|_{V'}\|\psi_x\|_V dt$$

$$+\frac{1}{2h}\int_{t_{i-1}}^{t_i}\|a\|_{1,\infty}\|w_i\|_{V'}^2 dt + \frac{1}{h}\int_{t_{i-1}}^{t_i}\|w_i\|_{V'}h^{1/4}\left\|h^{1/4}y_i^h\right\|_V dt.$$

Now for $t \in (t_{i-1}, t_i)$ we have

$$\|y(t_i) - y(t)\|_{V'} \le |t_i - t|^{1/2}\left\|\frac{dy}{dt}(t)\right\|_{L^2(0,T;V')} \le h^{1/2}C(\alpha_m). \qquad (2.95)$$

Here we recalled (2.55) which is inherited by its limit, too. Also,

$$\|a(y(t_i) - y(t))\|_{V'} \le \|a\|_{1,\infty}\|y(t_i) - y(t)\|_{V'} \text{ for } a \in W^{1,\infty}(\Omega).$$

We plug these two inequalities in (2.94) and performing some computations, we get that

$$\frac{\|w_i\|_{V'}^2}{2h} - \frac{\|w_{i-1}\|_{V'}^2}{2h} \leq \frac{C(\alpha_m)\sqrt{h}}{h} \int_{t_{i-1}}^{t_i} \|\zeta(t) - \zeta_i^h\|_V \, dt \tag{2.96}$$

$$+ \frac{\|a\|_{1,\infty}}{h} \int_{t_{i-1}}^{t_i} \|y(t) - y(t_i)\|_{V'} \|w_i\| \, dt + \frac{1}{2h} \int_{t_{i-1}}^{t_i} \|a\|_{1,\infty} \|w_i\|_{V'}^2 \, dt$$

$$+ \frac{1}{2} \|w_i\|_{V'}^2 + \frac{1}{2} \sqrt{h} \left\| h^{1/4} y_i^h \right\|_V^2$$

$$\leq C(\alpha_m) \frac{\sqrt{h}}{h} \left\{ \left(h + \int_{t_{i-1}}^{t_i} \|\zeta(t)\|_V^2 \, dt + \int_{t_{i-1}}^{t_i} \|\zeta_i^h\|_V^2 \, dt \right) + \int_{t_{i-1}}^{t_i} \|w_i\| \, dt \right\}$$

$$+ \frac{1}{2} (\|a\|_{1,\infty} + 1) \|w_i\|_{V'}^2 + \frac{1}{2} \sqrt{h} \left\| h^{1/4} y_i^h \right\|_V^2 .$$

By (2.36) and (2.53) we recall that $\|w_i\| \leq 2 \operatorname{ess\,sup}_{t \in (0,T)} \|y(t)\| \leq C$. We sum up (2.96) with respect to $i = 1, \ldots, p$, and using (2.42) we obtain that

$$\|w_p\|_{V'}^2 - \|w_0\|_{V'}^2 \leq C\sqrt{h} + (1 + \|a\|_{1,\infty}) h \sum_{i=1}^{p} \|w_i\|_{V'}^2 .$$

We take into account that $w_0 = y(0) - y_0^h = 0$, so that the latter inequality implies, by Lemma 2.4 with $C_M = 1 + \|a\|_{1,\infty}$ and $C_0 = C\sqrt{h}$, that

$$\|w_p\|_{V'}^2 \leq C(\alpha_m)\sqrt{h}$$

i.e., (2.92) as claimed. This ends the proof. □

We recall once again (2.31), (2.43), so a sufficient condition that enables the scheme convergence is $n = \left[\frac{T}{h}\right] + 1$, $k > 1$, or

$$h < \min \left\{ \frac{1}{k^2} \frac{1}{a_0^4}, \frac{1}{1 + \|a\|_{1,\infty}} \right\}. \tag{2.97}$$

2.1.6 Numerical Results

In this section we present some numerical simulations intended to put into evidence the effects induced by the diffusivity vanishing in the subsets $Q_0 = \{(t, x); y(t, x) = 0\}$, the formation and advance of the free boundary between the saturated and unsaturated regions and the influence of the advection.

The numerical algorithm is constructed following the quasi m-accretivity proof. Instead of solving the system (2.23), with the multivalued A^h, we solve a system similar to (2.30). More specifically, denoting

$$\zeta_i^h \in \widetilde{\beta}^*(y_i^h), \quad \widetilde{\beta}^*(y_i^h) = \beta^*(y_i^h) + \sqrt{h}y_i^h, \quad K(G(\zeta_i^h)) = G(\zeta_i^h) \qquad (2.98)$$

we get the system

$$G^h(\zeta_i^h) + hA_1\zeta_i^h = hf_i^h + y_{i-1}^h, \quad i = 1,\ldots,n,$$

where

$$G^h(r) := \begin{cases} (\widetilde{\beta}^*)^{-1}(r) & \text{if } r < \beta_s^* + \sqrt{h}y_s \\ y_s & \text{if } r \geq \beta_s^* + \sqrt{h}y_s. \end{cases} \qquad (2.99)$$

After solving the system, we set

$$y_i^h := \begin{cases} (\widetilde{\beta}^*)^{-1}(\zeta_i^h) & \text{if } \zeta_i^h < \beta_s^* + \sqrt{h}y_s \\ y_s & \text{if } \zeta_i^h \geq \beta_s^* + \sqrt{h}y_s. \end{cases} \qquad (2.100)$$

We exemplify this method for a 2D process of water infiltration into a soil, considering a dimensionless degenerate fast diffusion model with advection, and a multivalued β^*,

$$\beta_{\mathrm{deg}}(r) = \frac{1}{2\sqrt{1-r}} - \frac{1}{2}, \ r \in [0,1), \ \beta_{\mathrm{deg}}^*(r) = \begin{cases} 1 - \sqrt{1-r} - \frac{r}{2}, & r \in [0,1) \\ \left[\frac{1}{2}, \infty\right), & r = 1, \end{cases} \qquad (2.101)$$

in the domain $\Omega = \{(x_1, x_2); x_1 \in (0,5), x_2 \in (0,5)\}$. The initial datum corresponds to a dry region Ω_0, the circle with center in $(2,3)$ and radius $\delta = 0.1$,

$$y_0(x_1, x_2) = \begin{cases} 0, & \text{on } \Omega_0 \\ \frac{(x_1-2)^2+(x_2-3)^2-0.1^2}{100}, & \text{otherwise} \end{cases} \qquad (2.102)$$

such that in this subset the diffusion coefficient β_{deg} vanishes. The other data are

$$K(r) = r, \ f = 0.1\exp(-x_1^2), \ \alpha = 0.00001, \ h = 0.2.$$

Simulations have been made using Comsol Multiphysics with Matlab for three cases with and without advection, and the solution is represented at times $t = 0.02, 1, 4.2, 10$.

In Fig. 2.1a–d the values of the solution y computed for the model without advection, $a = (0,0)$ are plotted. In Fig. 2.2a–d it is shown the solution

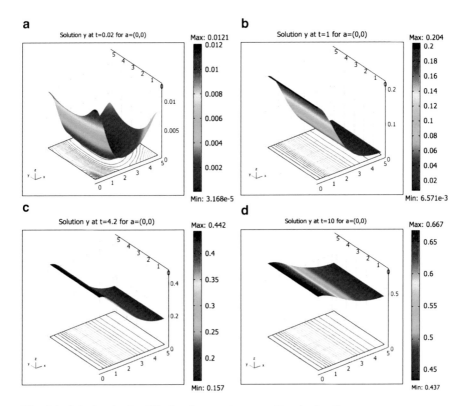

Fig. 2.1 Solution in the diffusion degenerate case without advection

y computed for the case with periodic advection components along Ox_1 and Ox_2, i.e., $a = (0.5 \sin \frac{\pi x_2}{30}, 0.5 \sin \frac{\pi x_1}{30})$ and in Fig. 2.3a–d we see the graphics of the solution in the case when the advection along Ox_2 has a larger amplitude, corresponding to $a = (0.5 \sin \frac{\pi x_2}{30}, 2.5 \sin \frac{\pi x_1}{30})$. It is observed how the vanishing diffusivity keeps the volumetric water content (or the moisture of the soil) y at low values in the domain Ω_0 at the beginning of the flow ($t = 0.02$) in all three cases. Then, at $t = 1$ the volumetric water content y remains at high values in a region in the neighborhood of $x = 0$ and this behavior is preserved at large time, too ($t = 10$) in the absence of advection in the first case. The cases with $a \neq 0$ reveal the influence of the variable advection which determines the water accumulation towards the corner $(5, 5)$ at intermediate times ($t = 4.2$) and even the saturation ($y = 1$) of this region due to the higher advection along Ox_2 in the third case. In Fig. 2.3c and d we notice the further advance of the saturated region along $x_2 = 5$ upwards. In the second case (Fig. 2.2) the increase of the volumetric water content in this region is done slowly, as a consequence of a slower advection.

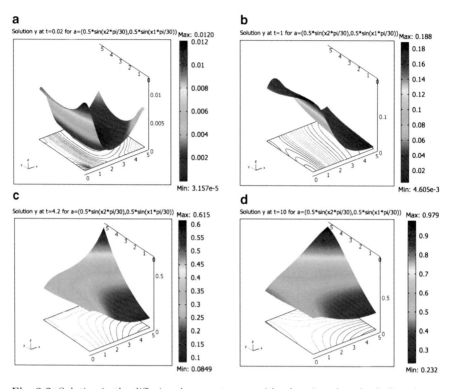

Fig. 2.2 Solution in the diffusion degenerate case with advection along both directions

2.2 Existence of Periodic Solutions in the Diffusion Degenerate Case

In this section we shall study the existence of periodic solutions to the degenerate diffusion problem without advection

$$\frac{\partial y}{\partial t} - \Delta \beta^*(y) \ni f \text{ in } \mathbb{R}_+ \times \Omega,$$

$$-\nabla \beta^*(y) \cdot \nu - \alpha \beta^*(y) \ni 0 \text{ on } \mathbb{R}_+ \times \Gamma, \tag{2.103}$$

$$y(t, x) = y(t + T, x) \text{ in } \Omega, \ t > 0$$

under the hypotheses

$$f \in C(\mathbb{R}_+; L^2(\Omega)), \ f(t + T, x) = f(t, x), \ f(t, x) \geq 0, \ \forall t \in \mathbb{R}_+, \text{ a.e. } x \in \Omega. \tag{2.104}$$

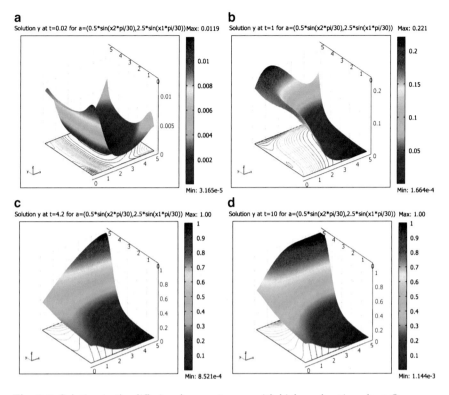

Fig. 2.3 Solution in the diffusion degenerate case with higher advection along Ox_2

We consider (2.4) with even powers $\beta(r) \geq \gamma_\beta r^{2m}$, for $r < 0$, $m \geq 1$, and in this case

$$j(r) \geq \frac{\gamma_\beta \, |r|^{2(m+1)}}{(2m+1)(2m+2)} \qquad (2.105)$$

(see (2.49)), hence

$$\lim_{|r| \to \infty} \frac{j(r)}{|r|} = +\infty. \qquad (2.106)$$

The study of this model on $(0, T)$ is done in the same functional framework as in Sect. 2.1 but the proof arguments will differ. We consider the problem on the period $(0, T)$,

$$\frac{\partial y}{\partial t} - \Delta \beta^*(y) \ni f \text{ in } Q = (0, T) \times \Omega,$$

$$-\nabla \beta^*(y) \cdot \nu - \alpha \beta^*(y) \ni 0 \text{ on } \Sigma = (0, T) \times \Gamma, \tag{2.107}$$

$$y(0, x) = y(T, x) \text{ in } \Omega.$$

Theorem 2.9. *Let us assume*

$$y_0 \in L^2(\Omega), \quad 0 \le y_0 \le y_s,$$

and let (2.104)–(2.105) hold. Then, there exists at least a nonnegative solution (y, ζ) to problem (2.103),

$$y \in C(\mathbb{R}_+; L^2(\Omega)) \cap W_{loc}^{1,2}(\mathbb{R}_+; V'),$$

$$\zeta \in L_{loc}^2(\mathbb{R}_+; V), \ \zeta(t, x) \in \beta^*(y(t, x)) \text{ a.e. } (t, x), \tag{2.108}$$

$$0 \le y(t, x) \le y_s, \text{ a.e. } (t, x) \in \mathbb{R}_+ \times \Omega.$$

Proof. We begin with the proof of the existence of a solution to (2.107). In problem (2.107) let us fix $y(0, x) = y_0$ in $L^2(\Omega)$ and consider the Cauchy problem

$$\frac{dy}{dt}(t) + Ay(t) \ni f(t), \text{ a.e. } t \in (0, T), \tag{2.109}$$

$$y(0) = y_0, \tag{2.110}$$

where A is given by (2.19) with $K_0 \equiv 0$. We prove that it has a unique solution, based on the fact that A is m-accretive on V'. Next, we shall define the mapping $\Phi : M_0 \to M_0$, $\Phi(y_0) = y(T)$, where

$$M_0 = \{v \in L^2(\Omega); 0 \le v(x) \le y_s \text{ a.e. } x \in \Omega\}$$

and prove that it satisfies the Schauder–Tikhonov theorem ([67], pp. 148).

We are going to prove the m-accretivity of A. The accretivity follows immediately by

$$(\xi - \bar{\xi}, y - \bar{y})_{V'} = \int_\Omega \nabla(\zeta - \bar{\zeta}) \cdot \nabla \psi dx = \int_\Omega (\zeta - \bar{\zeta})(y - \bar{y}) dx \ge 0,$$

where $\xi \in Ay$, $\bar{\xi} \in A\bar{y}$, $\zeta \in \beta^*(y)$, $\bar{\zeta} \in \beta^*(\bar{y})$ a.e. in Ω and ψ is the solution to $A_\Delta \psi = y - \bar{y}$ with A_Δ given by (2.16). For the m-accretivity we have to show that the equation

$$y + Ay = g \tag{2.111}$$

has a solution $y \in D(A)$ for each $g \in V'$.

We take $\zeta \in \beta^*(y)$ and denote $G(\zeta) = (\beta^*)^{-1}(\zeta)$. The function $G : \mathbb{R} \to (-\infty, y_s]$ is continuous and monotonically increasing (by the definition and properties of β^*), so it is maximal monotone on $\mathbb{R}$. Moreover, by (2.105) we have that

$$|G(\zeta)|^{2m+1} \leq C(|\zeta| + 1) \qquad (2.112)$$

with C a constant. Thus, we have to deal with the equation

$$G(\zeta) + A_0 \zeta = g, \qquad (2.113)$$

where A_0 is the restriction of A_Δ on $L^2(\Omega)$, $A_0 : D(A_0) \subset L^2(\Omega) \to L^2(\Omega)$,

$$D(A_0) = \left\{ v \in V; A_\Delta v \in L^2(\Omega), \frac{\partial v}{\partial \nu} + \alpha v = 0 \right\}, \quad A_0 v = -\Delta v.$$

First, we shall consider that $g \in L^2(\Omega)$. The linear operator $A_0 : V \to V'$ is continuous and coercive, so its realization on $L^2(\Omega)$ is m-accretive (see [14], pp. 36). Also, the realization of G on $L^2(\Omega)$ is maximal monotone on $L^2(\Omega)$. Hence, we define $G + A_0 : D(A + G_0) \subset L^2(\Omega) \to L^2(\Omega)$, where

$$D(G + A_0) = \{v \in D(A_0); G(\zeta) \in L^2(\Omega)\}$$

and show that

$$(A_0\zeta, G(\zeta)) = \int_\Omega \nabla\zeta \cdot \nabla G(\zeta) dx + \int_\Gamma \alpha \zeta G(\zeta) d\sigma \geq 0$$

which follows by the monotonicity of G and (2.5). We conclude by a known result (see [14], pp. 104) that $G + A_0$ is m-accretive on $L^2(\Omega)$. It is also coercive, so it is surjective (see [14], pp. 36) and therefore (2.113) has a unique solution $\zeta \in D(G + A_0)$.

Next, we fix $g \in V'$ and take a sequence $(g_n)_{n \geq 1} \subset L^2(\Omega)$, such that $g_n \to g$ in V'. Then

$$G(\zeta_n) + A_0\zeta_n = g_n \qquad (2.114)$$

will provide a sequence of solutions $(\zeta_n)_{n \geq 1}$ with $\zeta_n \in D(G + A_0)$ which satisfies the estimate

$$\|\zeta_n\|_V \leq C \|g_n\|_{V'}$$

obtained by multiplying (2.114) by ζ_n and using the fact that G is maximal monotone. Therefore, there exists a subsequence such that $\zeta_n \rightharpoonup \zeta$ in V and $\zeta_n \to \zeta$ in $L^2(\Omega)$ by the compactness of the injection of V in $L^2(\Omega)$.

By (2.112) we get that $G(\zeta_n)$ is bounded in $L^2(\Omega)$, so that $G(\zeta_n) \rightharpoonup \kappa$ in $L^2(\Omega)$ on a subsequence. By (2.114) it follows that $A_0\zeta_n$ is bounded in V', so that $A_0\zeta_n \rightharpoonup A_0\zeta$ in V'. Again by (2.114) we have that

$$\limsup_{n \to \infty} (G(\zeta_n), \zeta_n) \le (\kappa, \zeta).$$

Consequently, $\kappa = G(\zeta)$ a.e. in Ω (see [10], pp. 42) and passing to the limit in (2.114) we deduce that ζ is the solution to (2.113).

Because A is m-accretive and $K_0 = 0$ it follows by Theorem 2.6 that (2.109)–(2.110) has a unique solution y belonging to the spaces specified in (2.108).

Moreover, if $f \ge 0$ and $y_0 \ge 0$, the solution is nonnegative and this is proved by multiplying (2.109) by the negative part y^-, integrating over $(0, t)$ and applying the Stampacchia lemma (see [13], pp. 166). We recall that the negative part is the function $y^- = \min(0, -y)$. We get

$$-\frac{1}{2} \left\| y^-(t) \right\|^2 + \frac{1}{2} \left\| y^-(0) \right\|^2 - \int_Q \beta(y) \nabla y \cdot \nabla y^- \, dx dt = \int_Q f y^- \, dx dt,$$

whence taking into account that $y^-(0) = y_0^- = 0$ and $f \ge 0$ we obtain that $\left\| y^-(t) \right\|^2 = 0$ for any $t \in [0, T]$, so that $y(t) \ge 0$.

Now, we return to the proof of the conditions in Schauder–Tikhonov theorem. By Proposition 2.7, for $K_0 \equiv 0$ the solution y to (2.109)–(2.110) is unique so Φ is single-valued. It is obvious that M_0 is weakly compact in $L^2(\Omega)$ since it is closed and convex. Then, $\Phi(M_0) \subset M_0$ because for $y_0 \in M_0$ we have $\Phi(y_0) = y(T) \in [0, y_s]$, so that $\Phi(M_0)$ is weakly compact, too.

Finally, we are going to show that Φ is weakly compact in $L^2(\Omega)$. Let $(y_0^n)_{n \ge 1} \subset L^2(\Omega)$, such that $y_0^n \rightharpoonup y_0$ in $L^2(\Omega)$ and consider the Cauchy problem

$$\frac{dy_n}{dt}(t) + Ay_n(t) \ni f(t), \quad \text{a.e. } t \in (0, T), \tag{2.115}$$

$$y_n(0) = y_0^n.$$

Again by Theorem 2.6 this problem has a unique solution (y_n, ζ_n), $\zeta_n \in \beta^*(y_n)$ a.e. in Q, which satisfies the estimate

$$\int_\Omega j(y_n(x, t)) dx + \int_0^t \left\| \frac{dy_n}{d\tau}(\tau) \right\|_{V'}^2 d\tau + \int_0^t \| \zeta_n(\tau) \|_V^2 \, d\tau \le C(\alpha_m),$$

obtained by (2.61)–(2.63), with $C(\alpha_m)$ depending on $\alpha_m^{-1/2}$. This implies that we can select a subsequence such that

$$y_n \rightharpoonup y \text{ weakly in } L^2(0, T; L^2(\Omega)) \text{ as } n \to \infty,$$

$$\frac{dy_n}{dt} \rightharpoonup \frac{dy}{dt} \text{ in } L^2(0, T; V') \text{ as } n \to \infty,$$

$$Ay_n \ni \zeta_n \rightharpoonup \zeta \text{ in } L^2(0,T;V) \text{ as } n \to \infty,$$

$$y_n(t) \to y(t) \text{ in } V' \text{ as } n \to \infty,$$

$$y_n(T) \to y(T) \text{ in } V' \text{ as } n \to \infty,$$

$$y_n(0) \to y_0 \text{ in } V' \text{ as } n \to \infty.$$

Then we have

$$\int_0^T (\zeta_n(t), y_n(t))_{V'} dt = -\frac{1}{2} \|y_n(T)\|_{V'}^2 + \frac{1}{2} \|y_0^n\|_{V'}^2 + \int_0^T (f(t), y_n(t))_{V'} dt$$

whence

$$\limsup_{n \to \infty} \int_0^T (\zeta_n(t), y_n(t))_{V'} dt$$

$$= -\frac{1}{2} \|y(T)\|_{V'}^2 + \frac{1}{2} \|y_0\|_{V'}^2 + \int_0^T (f(t), y(t))_{V'} dt = \int_0^T (\zeta(t), y(t))_{V'} dt.$$

We deduce that $\zeta \in Ay$ a.e. in Q and conclude that y is the solution to (2.109)–(2.110). Moreover, $\Phi(y_0^n) = y^n(T) \rightharpoonup y(T) = \Phi(y_0)$ in $L^2(\Omega)$. Thus, Φ has a fixed point, $y(T) = \Phi(y_0) = y_0$ and this ends the proof.

The proof of a solution to (2.103) is done as in Theorem 1.13, starting by (2.107) and extending the solution to $\mathbb{R}_+$ by periodicity. □

2.2.1   Asymptotic Behavior at Large Time

In a Hilbert space H let A be a maximal monotone operator provided by a proper convex lower semicontinuous function φ, $A = \partial\varphi$. Let us consider the abstract equation

$$\frac{dy}{dt}(t) + Ay(t) \ni f(t) \text{ a.e. } t \in \mathbb{R}. \tag{2.116}$$

We recall the following general result (see [68]).

Theorem H. *Let A be a potential operator and let f be T-periodic, i.e.,*

$$f(t+T) = f(t) \text{ a.e. } t \in \mathbb{R}.$$

Assume that (2.116) has at least a T-periodic solution. Then, every other solution y, corresponding to a whatever initial datum y_0, is such that

$$y(t) - \pi(t) \to 0 \text{ as } t \to \infty, \tag{2.117}$$

where $\pi(t)$ is some periodic solution to (2.116). Two periodic solutions ω and π differ by a constant element of H.

Moreover, if the resolvent of A is compact in H, then the convergence (2.117) is strong.

Lemma 2.10. *If $K_0 = 0$ the operator A defined by (2.19) is a potential operator.*

Proof. We define $\varphi : V' \to (-\infty, \infty]$

$$\varphi(y) = \begin{cases} \int_\Omega j(y)dx, & \text{if } y \in V' \cap L^1(\Omega), \ j(y) \in L^1(\Omega) \\ +\infty, & \text{otherwise} \end{cases}$$

and notice that $D(\varphi) = \{y \in L^2(\Omega); y \leq y_s \text{ a.e. } x \in \Omega\}$. Indeed, let $y \in L^2(\Omega)$, $y \leq y_s$ a.e. $x \in \Omega$. Then, $\varphi(y) \leq \varphi(y_s) = \beta_s^* y_s \text{meas}(\Omega) < \infty$, hence $y \in D(\varphi)$. The converse inclusion is implied by the inequality (2.105). If $y \in D(\varphi)$ we have

$$\|y\|^{2(m+1)} \leq C(2m+1)(2m+2)\varphi(y) < +\infty \tag{2.118}$$

whence we obtain that $y \in L^2(\Omega)$. Moreover, the fact that $j(y) \in L^1(\Omega)$ implies $y \leq y_s$ a.e. $x \in \Omega$.

Then we show that φ is a proper, convex and lower semicontinuous (l.s.c.). The argument is well known (see [10]) but we outline it for reader's convenience. The first two assertions are obvious and it remains to show that the set

$$S = \{y \in V'; \varphi(y) \leq \lambda\}$$

is closed in V' for any $\lambda > 0$.

First we show that S is strongly closed in $L^2(\Omega)$. Indeed, let $y_n \in S$, $y_n \to y$ in $L^2(\Omega)$. Recalling that j is l.s.c. and nonnegative we have by Fatou's lemma that

$$\varphi(y) = \int_\Omega j(y)dx \leq \int_\Omega \liminf_{n\to\infty} j(y_n)dx \leq \liminf_{n\to\infty} \int_\Omega j(y_n)dx,$$

which implies that $\varphi(y) \leq \lambda$, i.e., $y \in S$.

Now we can prove that S is closed in V'. Let $(y_n)_n$ be a sequence in S such that $y_n \to y$ in V' as $n \to \infty$. We have to show that $y \in S$, meaning that $\varphi(y) \leq \lambda$.

By (2.118) it follows that $(y_n)_n$ is bounded in $L^2(\Omega)$, so that we can extract a subsequence such that

$$y_n \rightharpoonup \overline{y} \text{ in } L^2(\Omega) \text{ as } n \to \infty. \tag{2.119}$$

By the uniqueness of the limit ($y_n \to y$ in V') it follows that $\bar{y} = y$. Because S is convex and strongly closed in $L^2(\Omega)$ it follows that it is weakly closed too, whence we get that $y \in S$.

Finally we shall show that $A = \partial \varphi$. First we compute

$$(\xi, y - \bar{y})_{V'} = \int_\Omega \nabla \zeta \cdot \nabla \psi dx + \int_\Gamma \alpha \zeta \psi d\sigma = \int_\Omega \zeta (y - \bar{y}) dx \text{ for any } y, \bar{y} \in V,$$

where $\xi \in Ay$ a.e. $x \in \Omega$, $\zeta \in \beta^*(y)$ a.e. on Ω and $\psi \in V$ satisfies $A_\Delta \psi = y - \bar{y}$. But $\partial j(y) = \beta^*(y)$ so that

$$(\xi, y - \bar{y})_{V'} \geq \int_\Omega (j(y) - j(\bar{y})) dx, \ \xi \in Ay \text{ a.e. on } \Omega,$$

which shows that $Ay \subset \partial \varphi(y)$. Since A is maximal monotone in $V' \times V'$ it follows that $A = \partial \varphi$. $\qquad \square$

Now we shall give a result showing that the solution to the problem

$$\frac{\partial y}{\partial t} - \Delta \beta^*(y) \ni f \text{ in } \mathbb{R}_+ \times \Omega,$$

$$-\nabla \beta^*(y) \cdot \nu - \alpha \beta^*(y) \ni 0 \text{ on } \mathbb{R}_+ \times \Gamma, \qquad (2.120)$$

$$y(0) = y_0$$

starting from y_0 tends in some sense to a solution to (2.103).

Corollary 2.11. *The solution y to (2.120) corresponding to a whatever datum y_0 satisfies*

$$\lim_{t \to \infty} \|y(t) - \omega(t)\|_{V'} = 0, \qquad (2.121)$$

where ω is some periodic solution to (2.103).

Proof. The operator A is potential and (2.103) has at least a periodic solution.

Next we will show that $(\lambda I + A)^{-1}$ for $\lambda > 0$ is compact in V'. Let $(v_n)_n$ be a sequence in V'. We recall that A is m-accretive on V' and note that it is coercive too, because $(\xi, y)_{V'} \geq \rho \|y\|^2 \geq \frac{\rho}{c_H^2} \|y\|_{V'}^2$, where $\xi \in Ay$ and c_H is given by (2.14). Hence A is surjective and the equation

$$u_n + Au_n = v_n,$$

has a unique solution in $D(A)$. We multiply it by $\zeta_n \in \beta^*(u_n)$ and integrate over Ω. We get that

$$\|u_n\|^2 + \|\zeta_n\|_V^2 \leq \|v_n\|_{V'}^2 \leq \text{constant.}$$

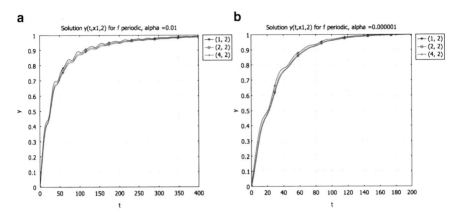

Fig. 2.4 Asymptotic behavior of $\theta = uy$ solution to (2.120) in the periodic diffusion degenerate case

Since $L^2(\Omega)$ is compact in V' it follows that $(u_n)_n$ is compact in V', where $u_n = (\lambda I + A)^{-1}v_n$. Then, Theorem H implies (2.121).   □

2.2.2   Numerical Results

The numerical results presented below are intended to illustrate the result proved in Corollary 2.11.

We perform the computations for $\Omega = (0,2) \times (0,5)$, Ω_0 the circle with center at $(2,3)$ and radius 0.1, β^* given by (2.101), y_0 given by (2.102), $K(r) = 0$ and

$$f(t, x_1, x_2) = 0.001x_1(5 - x_2)\left(\left|\sin\frac{\pi}{20}t\right| + \left|\cos\frac{\pi}{30}t\right|\right).$$

We shall also put into evidence the influence of the boundary condition upon the flow. In Fig. 2.4a the values are computed at $x_2 = 2$, $x_1 = 1, 2, 4$, for $\alpha = 0.01$ meaning that the boundary of the domain is permeable and allows the fluid flux across it. In Fig. 2.4b there are the graphics computed for $\alpha = 0.000001$ which indicates an almost impermeable boundary. In both cases the flow ends by completely filling the soil reaching so a stationary regime, but this is done in a shorter time in the second case when the flux across the boundary is very much reduced. In the first case the flow needs a double time to reach the (periodic) stationary regime than in the first case.

Chapter 3
Existence for Nonautonomous Parabolic–Elliptic Degenerate Diffusion Equations

The subject of this chapter is the study of a Cauchy problem with a time-dependent nonlinear operator which is the abstract formulation of a boundary value problem for a fast diffusion equation in the parabolic–elliptic degenerate case, with nonhomogeneous Neumann conditions. Existence and uniqueness for the abstract Cauchy problem are proved in relation with the results of Kato, given in [71] and extended by Crandall and Pazy in [44] for the nonautonomous evolution equations with nonlinear m-accretive operators.

3.1 Statement of the Problem and Functional Framework

Let Ω be an open bounded subset of $\mathbb{R}^N$, with a piecewise smooth boundary $\Gamma := \partial\Omega$. We assume that $\Gamma = \Gamma_p \cup \overline{\Gamma_\alpha}$, where Γ_p and Γ_α are disjoint smooth open subsets of the boundary and denote by ν the unit outward normal to the boundary Γ. We are concerned with the boundary value problem with initial data

$$\frac{\partial(u(t,x)y)}{\partial t} - \Delta\beta^*(y) + \nabla \cdot K_0(t,x,y) \ni f, \qquad \text{in } Q = (0,T) \times \Omega, \quad (3.1)$$

$$(u(t,x)y(t,x))|_{t=0} = \theta_0, \qquad \text{in } \Omega,$$

$$(K_0(t,x,y) - \nabla\beta^*(y)) \cdot \nu \ni p(t,x), \quad \text{on } \Sigma_p = (0,T) \times \Gamma_p,$$

$$y(t,x) = 0, \qquad \text{on } \Sigma_\alpha = (0,T) \times \Gamma_\alpha.$$

The hypotheses for β and β^* are the same as in Sect. 1.1, i.e., (1.2)–(1.9). For u we assume

A. Favini and G. Marinoschi, *Degenerate Nonlinear Diffusion Equations*,
Lecture Notes in Mathematics 2049, DOI 10.1007/978-3-642-28285-0_3,
© Springer-Verlag Berlin Heidelberg 2012

$$u \in W^{2,\infty}(Q), \quad u(t,x) \in [0, u_M] \text{ for any } (t,x) \in Q,$$

$$u(0,x) := u_0(x), \tag{3.2}$$

$$u(t,x) = 0 \text{ in } \overline{Q_0} = [0,T] \times \overline{\Omega_0}, \quad u(t,x) > 0 \text{ on } Q_u = (0,T) \times \Omega_u,$$

where $\overline{\Omega_0} \subset \Omega$, $\text{meas}(\Omega_0) > 0$, $\overline{\Omega_0} \cap \Gamma = \varnothing$, $\Omega_u = \Omega \backslash \overline{\Omega_0}$.

In this chapter we consider that the advection term is time-dependent

$$K_0(t,x,y) = a(t,x)K(y), \quad a(t,x) := (a_j(t,x))_{j=1,\dots,N},$$

with a and K satisfying

$$a_j \in W^{1,\infty}(Q), \ a_j(t,x) = 0 \text{ in } \overline{Q_0}, \ |a_j(t,x)| \leq a_j^M, \text{ for } (t,x) \in \overline{Q}, \tag{3.3}$$

the function $K : (-\infty, y_s] \to \mathbb{R}$ is Lipschitz continuous with the constant $M_K > 0$, and bounded, $|K(r)| \leq K_s$.

We also assume

$$f \in L^2(0,T;V'), \quad p \in L^2(0,T;L^2(\Gamma_p))$$

and

$$\theta_0 \in L^2(\Omega), \ \theta_0 = 0 \text{ a.e. on } \Omega_0, \ \theta_0 \geq 0 \text{ a.e. on } \Omega_u, \tag{3.4}$$

$$\frac{\theta_0}{u_0} \in L^2(\Omega_u), \ \frac{\theta_0}{u_0} \leq y_s, \text{ a.e. } x \in \Omega_u.$$

We consider the Hilbert space $V := \{y \in H^1(\Omega); \ y = 0 \text{ on } \Gamma_\alpha\}$ with the norm

$$\|y\|_V = \left(\int_\Omega |\nabla y(x)|^2 \, dx \right)^{1/2}$$

and its dual, V'. We endow V' with the scalar product $(y, \overline{y})_{V'} = \langle y, \psi \rangle_{V',V}$ where ψ is the solution to the problem

$$-\Delta\psi = \overline{y}, \ \psi = 0 \text{ on } \Gamma_\alpha, \ \frac{\partial\psi}{\partial\nu} = 0 \text{ on } \Gamma_p$$

which can still be expressed as $A_0\psi = \overline{y}$ with $A_0 : V \to V'$ defined by

$$\langle A_0\psi, \phi \rangle_{V',V} := \int_\Omega \nabla\psi \cdot \nabla\phi \, dx, \text{ for any } \phi \in V. \tag{3.5}$$

We recall that for the sake of simplicity we shall denote the norm and scalar product in $L^2(\Omega)$ without subscript.

Definition 3.1. We call a *weak solution* to (3.1) a pair (y, ζ) such that

$$y \in L^2(0, T; V), \tag{3.6}$$

$$uy \in C([0, T]; L^2(\Omega)) \cap L^2(0, T; V) \cap W^{1,2}([0, T]; V'),$$

$$\zeta \in L^2(0, T; V), \ \zeta(t, x) \in \beta^*(y(t, x)) \text{ a.e. on } Q,$$

$$y \le y_s, \text{ a.e. } (t, x) \in Q,$$

which satisfies the equation

$$\left\langle \frac{d(uy)}{dt}(t), \psi \right\rangle_{V', V} + \int_\Omega (\nabla \zeta(t) \cdot \nabla \psi - K_0(t, x, y(t)) \cdot \nabla \psi) \, dx \tag{3.7}$$

$$= \langle f(t), \psi \rangle_{V', V} - \int_{\Gamma_p} p(t)\psi dx, \text{ a.e. } t \in (0, T), \ \forall \psi \in V,$$

and the initial condition $(u(t)y(t))|_{t=0} = \theta_0$.

Equivalently (3.7) can be written

$$\int_0^T \left\langle \frac{d(uy)}{dt}(t), \phi(t) \right\rangle_{V', V} dt + \int_Q (\nabla \zeta \cdot \nabla \phi - K_0(t, x, y) \cdot \nabla \phi) \, dx dt$$

$$= \int_0^T \langle f(t), \phi(t) \rangle_{V', V} \, dt - \int_{\Sigma_p} p\phi d\sigma dt, \tag{3.8}$$

for any $\phi \in L^2(0, T; V)$, and some $\zeta(t, x) \in \beta^*(y(t, x))$ a.e. $(t, x) \in Q$.
We replace

$$\theta(t, x) := u(t, x)y(t, x), \tag{3.9}$$

and for each $t \in [0, T]$ we introduce the operator $A(t) : D(A(t)) \subset V' \to V'$, with

$$D(A(t)) \tag{3.10}$$

$$:= \left\{ z \in L^2(\Omega); \ \frac{z}{u(t, \cdot)} \in L^2(\Omega), \ \exists \zeta \in V, \ \zeta(x) \in \beta^*\left(\frac{z(x)}{u(t, x)}\right) \text{ a.e. } x \in \Omega \right\},$$

by the relation

$$\langle A(t)z, \psi \rangle_{V', V} := \int_\Omega \left(\nabla \zeta \cdot \nabla \psi - K_0\left(t, x, \frac{z}{u(t, x)}\right) \cdot \nabla \psi \right) dx, \tag{3.11}$$

for any $\psi \in V$.

If $p \in L^2(0, T; L^2(\Gamma_p))$ we define $f_{\Gamma_p} \in L^2(0, T; V')$ by

$$\langle f_{\Gamma_p}(t), \psi \rangle_{V', V} := - \int_{\Gamma_p} p(t) \psi d\sigma, \quad \forall \psi \in V, \tag{3.12}$$

and so we are led to the Cauchy problem

$$\frac{d\theta}{dt}(t) + A(t)\theta(t) \ni f(t) + f_{\Gamma_p}(t), \quad \text{a.e. } t \in (0, T), \tag{3.13}$$

$$\theta(0) = \theta_0.$$

We notice that if the solution to (3.13) exists, then it is a solution in the generalized sense to (3.1).

3.2   The Approximating Problem

We introduce an approximating problem by replacing u by $u_\varepsilon = u + \varepsilon$, the multivalued function β^* by the continuous approximation β^*_ε defined in (1.36) and extend $K(y)$ at the right of $y = y_s$ by $K(y_s)$.

Therefore, we write the Cauchy problem

$$\frac{d\theta_\varepsilon}{dt}(t) + A_\varepsilon(t)\theta_\varepsilon(t) = f(t) + f_{\Gamma_p}(t), \quad \text{a.e. } t \in (0, T), \tag{3.14}$$

$$\theta_\varepsilon(0) = \theta_0,$$

with the time-dependent operator $A_\varepsilon(t) : D(A_\varepsilon(t)) \subset V' \to V'$ defined by

$$\langle A_\varepsilon(t)z, \psi \rangle_{V', V} \tag{3.15}$$

$$:= \int_\Omega \left(\nabla \beta^*_\varepsilon \left(\frac{z(x)}{u_\varepsilon(t, x)} \right) \cdot \nabla \psi - K_0 \left(t, x, \frac{z(x)}{u_\varepsilon(t, x)} \right) \cdot \nabla \psi \right) dx, \quad \forall \psi \in V$$

for each $t \in [0, T]$ with

$$D(A_\varepsilon(t)) := \left\{ z \in L^2(\Omega); \ \beta^*_\varepsilon \left(\frac{z}{u_\varepsilon(t, \cdot)} \right) \in V \right\}.$$

Equivalently to (3.14) we can write

$$\int_0^T \left\langle \frac{d\theta_\varepsilon}{dt}(t), \phi(t) \right\rangle_{V',V} dt$$

$$+ \int_Q \left(\nabla \beta_\varepsilon^* \left(\frac{\theta_\varepsilon}{u_\varepsilon} \right) \cdot \nabla \phi - K_0 \left(t, x, \frac{\theta_\varepsilon}{u_\varepsilon} \right) \cdot \nabla \phi \right) dx dt \qquad (3.16)$$

$$= \int_0^T \langle f(t), \phi(t) \rangle_{V',V} dt - \int_{\Sigma_p} p \phi d\sigma dt, \ \forall \phi \in L^2(0,T;V).$$

The definitions and properties of the functions j and j_ε are the same as in Sect. 1.1.

Since (3.14) is a nonautonomous problem a way of solving it would be via Lions' theorem. For that the operator $A_\varepsilon(t)$ should be defined from V to V' and must be monotone. But this is not the case of this operator in the duality V', V. That is why we have to find another way of treating this problem and approach it relying on the results of Kato [71], extended by Crandall and Pazy in [44].

First we prove a lemma that gathers some important properties of $A_\varepsilon(t)$ which are the hypotheses assumed in [44].

Lemma 3.2. *Let $A_\varepsilon(t)$ be the operator defined by (3.15).*

(a) *The domain of $A_\varepsilon(t)$ is independent of t and $D(A_\varepsilon(t)) = D(A_\varepsilon(0)) = V$.*
(b) *For each $\varepsilon > 0$ and $t \in [0,T]$ fixed, the operator $A_\varepsilon(t)$ is quasi m-accretive on V'.*
(c) *For $\theta \in V$ and $0 \le s, t \le T$ we have*

$$\|A_\varepsilon(t)\theta - A_\varepsilon(s)\theta\|_{V'} \le |t - s| \, g\left(\|\theta\|_{V'}\right) \left(\|A(t)\theta\|_{V'} + 1\right), \qquad (3.17)$$

where $g : [0, \infty) \to [0, \infty)$ is a nondecreasing function.

Proof. (a) If $\theta \in V$ then $\frac{\theta}{u_\varepsilon} \in V$, since $u_\varepsilon \in u \in W^{2,\infty}(Q)$ and u_ε is in the bounded interval $[\varepsilon, u_M + \varepsilon]$. Next, we easily see that $\frac{\partial}{\partial x_j}\left(\beta_\varepsilon^*\left(\frac{\theta}{u_\varepsilon}\right)\right) = \beta_\varepsilon\left(\frac{\theta}{u_\varepsilon}\right)\frac{\partial}{\partial x_j}\left(\frac{\theta}{u_\varepsilon}\right) \in L^2(\Omega)$ for each $\varepsilon > 0$. Thus, $V \subset D(A_\varepsilon(t))$. Conversely, since the inverse of β_ε^* is Lipschitz with the constant $\frac{1}{\rho}$, it follows that $\beta_\varepsilon^*\left(\frac{\theta}{u_\varepsilon}\right) \in V$ implies $\frac{\theta}{u_\varepsilon} \in V$ and finally $\theta \in V$, whence $D(A_\varepsilon(t)) \subset V$, for each $t \in [0,T]$.

(b) Let $\lambda > 0$, and $t \in [0,T]$ be fixed. The quasi m-accretivity of $A_\varepsilon(t)$ follows as in Lemma 1.4 (for B_ε).

To prove (c) we calculate for any $t, s, 0 \le s < t \le T$

$$\|A_\varepsilon(t)\theta - A_\varepsilon(s)\theta\|_{V'}^2 = \langle A_\varepsilon(t)\theta - A_\varepsilon(s)\theta, \psi \rangle_{V',V}, \ \text{for } \theta \in V,$$

where
$$-\Delta\psi = A_\varepsilon(t)\theta - A_\varepsilon(s)\theta, \quad \frac{\partial\psi}{\partial\nu} = 0 \text{ on } \Gamma_p, \ \psi = 0 \text{ on } \Gamma_\alpha.$$

By hypotheses (3.2) and (3.3) the functions $\frac{1}{u_\varepsilon(t,x)}$, $\nabla\left(\frac{1}{u_\varepsilon(t,x)}\right)$ and a_j are Lipschitz continuous with respect to t, uniformly with respect to x, i.e.,

$$\left|\frac{1}{u_\varepsilon(t,x)} - \frac{1}{u_\varepsilon(s,x)}\right| \leq C(\varepsilon)\,|t-s|\,, \forall x \in \Omega, \tag{3.18}$$

$$\left|\nabla\left(\frac{1}{u_\varepsilon(t,x)}\right) - \nabla\left(\frac{1}{u_\varepsilon(s,x)}\right)\right| \leq C(\varepsilon)\,|t-s|\,, \forall x \in \Omega. \tag{3.19}$$

Using (3.3) and the hypotheses for K we can write

$$\int_\Omega \left[K_0\left(t,x,\frac{\theta}{u_\varepsilon(t,x)}\right) - K_0\left(t,x,\frac{\theta}{u_\varepsilon(s,x)}\right)\right] \cdot \nabla\psi dx$$

$$\leq \left\|(a(t,\cdot) - a(s,\cdot))K\left(t,x,\frac{\theta}{u_\varepsilon(t,\cdot)}\right)\right\| \|\nabla\psi\|$$

$$+ \left\|a(s,\cdot)\left(K\left(t,x,\frac{\theta}{u_\varepsilon(t,\cdot)}\right) - K\left(t,x,\frac{\theta}{u_\varepsilon(s,\cdot)}\right)\right)\right\| \|\nabla\psi\|$$

$$\leq \left(K_s C_a\,|t-s| + \overline{M}\left\|\frac{\theta}{u_\varepsilon(t,\cdot)} - \frac{\theta}{u_\varepsilon(s,\cdot)}\right\|\right) \|\nabla\psi\|.$$

By $C(\varepsilon)$ we denote several constants depending on ε and C_a is the Lipschitz constant for a. Therefore, taking account of the definition of β_ε^* from (1.36) we have that

$$\|A_\varepsilon(t)\theta - A_\varepsilon(s)\theta\|_{V'}^2 = \int_\Omega \left\{\nabla\left[\beta_\varepsilon^*\left(\frac{\theta}{u_\varepsilon(t,x)}\right) - \beta_\varepsilon^*\left(\frac{\theta}{u_\varepsilon(s,x)}\right)\right] \cdot \nabla\psi\right.$$

$$\left. - \left[K_0\left(t,x,\frac{\theta}{u_\varepsilon(t,x)}\right) - K_0\left(t,x,\frac{\theta}{u_\varepsilon(s,x)}\right)\right] \cdot \nabla\psi\right\} dx$$

$$\leq \left|\left\langle A_\varepsilon(t)\theta - A_\varepsilon(s)\theta, \beta_\varepsilon^*\left(\frac{\theta}{u_\varepsilon(t,x)}\right) - \beta_\varepsilon^*\left(\frac{\theta}{u_\varepsilon(s,x)}\right)\right\rangle_{V',V}\right|$$

$$+ \left(K_s C_a\,|t-s| + \overline{M}\left\|\frac{\theta}{u_\varepsilon(t,\cdot)} - \frac{\theta}{u_\varepsilon(s,\cdot)}\right\|\right) \|\psi\|_V$$

$$\leq \frac{\beta_s^* - \beta^*(y_s - \varepsilon)}{\varepsilon} \|A_\varepsilon(t)\theta - A_\varepsilon(s)\theta\|_{V'} \left\|\frac{\theta}{u_\varepsilon(t,\cdot)} - \frac{\theta}{u_\varepsilon(s,\cdot)}\right\|_V$$

$$+ \left(K_s C_a\,|t-s| + \overline{M}c_P\left\|\frac{\theta}{u_\varepsilon(t,\cdot)} - \frac{\theta}{u_\varepsilon(s,\cdot)}\right\|_V\right) \|A_\varepsilon(t)\theta - A_\varepsilon(s)\theta\|_{V'},$$

where $\overline{M} = M_K \sum_{j=1}^{N} a_j^M$. Using (3.18) and (3.19) and performing some computations we deduce the estimate

$$\left\| \theta \left(\frac{1}{u_\varepsilon(t,\cdot)} - \frac{1}{u_\varepsilon(s,\cdot)} \right) \right\|_V \leq C(\varepsilon) \, |t-s| \, \|\theta\|_V \,,$$

so that we get that

$$\|A_\varepsilon(t)\theta - A_\varepsilon(s)\theta\|_{V'} \leq (K_s C_a + C(\varepsilon) \|\theta\|_V) \, |t-s| \,. \qquad (3.20)$$

We write $\theta = \frac{\theta}{u_\varepsilon} u_\varepsilon$ and compute $\|\theta\|_V$ obtaining

$$\|\theta\|_V \leq C \left\| \frac{\theta}{u_\varepsilon(t,\cdot)} \right\|_V, \qquad (3.21)$$

with C a constant depending on Ω and the norm $\|u\|_{C^1(Q)}$. Then, we compute

$$\left\langle A_\varepsilon(t)\theta, \frac{\theta}{u_\varepsilon(t,\cdot)} \right\rangle_{V',V}$$
$$= \int_\Omega \left(\nabla \beta_\varepsilon^* \left(\frac{\theta}{u_\varepsilon} \right) \cdot \nabla \left(\frac{\theta}{u_\varepsilon} \right) - K_0 \left(t, x, \frac{\theta}{u_\varepsilon} \right) \cdot \nabla \left(\frac{\theta}{u_\varepsilon} \right) \right) dx$$
$$\geq \int_\Omega \beta_\varepsilon \left(\frac{\theta}{u_\varepsilon(t,x)} \right) \left| \nabla \left(\frac{\theta}{u_\varepsilon(t,x)} \right) \right|^2 dx - M_1 \left\| \frac{\theta}{u_\varepsilon(t,\cdot)} \right\|_V$$
$$\geq \rho \left\| \frac{\theta}{u_\varepsilon(t,\cdot)} \right\|_V^2 - M_1 \left\| \frac{\theta}{u_\varepsilon(t,\cdot)} \right\|_V \,,$$

with $M_1 = K_s \sum_{j=1}^{N} a_j^M$, where we used the boundedness of K, $|K(r)| \leq K_s$. We deduce that

$$\rho \left\| \frac{\theta}{u_\varepsilon(t,\cdot)} \right\|_V^2 \leq \left\langle A_\varepsilon(t)\theta, \frac{\theta}{u_\varepsilon(t,\cdot)} \right\rangle_{V',V} + M_1 \left\| \frac{\theta}{u_\varepsilon(t,\cdot)} \right\|_V$$
$$\leq (\|A_\varepsilon(t)\theta\|_{V'} + M_1) \left\| \frac{\theta}{u_\varepsilon(t,\cdot)} \right\|_V$$

whence we get

$$\left\| \frac{\theta}{u_\varepsilon(t,\cdot)} \right\|_V \leq \frac{\max\{1, M_1\}}{\rho} \, (\|A_\varepsilon(t)\theta\|_{V'} + 1) \,. \qquad (3.22)$$

Combining (3.20)–(3.22) we finally obtain that

$$\|A_\varepsilon(t)\theta - A_\varepsilon(s)\theta\|_{V'} \le C(\varepsilon)\,|t - s|\,(\|A_\varepsilon(t)\theta\|_{V'} + 1),$$

as claimed by (3.17).   □

3.3   Well-Posedness for the Nonautonomous Cauchy Problem

Now we can pass to the proof of the existence and uniqueness results for problem (3.13).

We recall that by c_P we denote the constant in the Poincaré inequality and let $u'_m, u'_M \in \mathbb{R}$,

$$u'_M := \max_{(t,x)\in\overline{Q}} u_t(t,x), \quad u'_m := \min_{(t,x)\in\overline{Q}} u_t(t,x).$$

Theorem 3.3. *Let us assume* $f \in L^2(0,T;V')$, $p \in L^2(0,T;L^2(\Gamma_p))$, *(3.4) and*

$$\rho - c_P^2 u'_M > 0. \tag{3.23}$$

Then, the Cauchy problem (3.13) has a solution (y,ζ)

$$y \in L^2(0,T;V), \tag{3.24}$$

$$uy \in C([0,T];L^2(\Omega)) \cap L^2(0,T;V) \cap W^{1,2}([0,T];V')$$

$$\zeta \in L^2(0,T;V), \ \zeta \in \beta^*(y) \ a.e. \ on \ Q,$$

$$y(t,x) \le y_s \ a.e. \ (t,x) \in Q.$$

If there exists $k_u > 0$ *such that*

$$|a(t,x)| \le k_u \sqrt{u(t,x)} \ for \ all \ (t,x) \in Q$$

then the solution is unique.

Proof. Since the result of Crandall and Pazy we intend to use works for homogeneous evolution equations we first resort to a technique developed by Dafermos and Slemrod in [47] by which (3.14) is transformed into a homogeneous problem. For the homogeneous problem we apply then the main theorem stating the existence of solutions to evolution equations with time-dependent nonlinear operators given in [44] after we prove that all hypotheses specified there are fulfilled.

The proof is done in three steps. In the first two steps we show that the approximating homogeneous problem has a unique solution and then we pass to the limit as $\varepsilon \to 0$ for getting the existence for (3.13).

Step 1. First, let us assume that

$$f \in W^{1,1}(0, \infty; V'), \ f_{\Gamma_p} \in W^{1,1}(0, \infty; V'), \ \theta_0 \in V. \qquad (3.25)$$

We denote $F = f + f_{\Gamma_p} \in W^{1,1}(0, \infty; V')$ and introduce the space $\mathcal{X} := V' \times L^1(0, \infty; V')$ with the norm

$$\|\Theta\|_{\mathcal{X}}^2 = \|\theta\|_{V'}^2 + \int_0^\infty \|\gamma(s)\|_{V'} \, ds, \text{ where } \Theta = \begin{pmatrix} \theta \\ \gamma \end{pmatrix} \in \mathcal{X}.$$

Then we define the operator

$$\mathcal{A}_\varepsilon(t) : D(\mathcal{A}_\varepsilon(t)) = D(A_\varepsilon(t)) \times W^{1,1}(0, \infty; V') \subset \mathcal{X} \to \mathcal{X}$$

by

$$\mathcal{A}_\varepsilon(t)(w, \gamma) = \begin{pmatrix} A_\varepsilon(t)w - \gamma(0) \\ -\gamma' \end{pmatrix}, \qquad (3.26)$$

for all $(w, \gamma) \in D(A_\varepsilon(t)) \times W^{1,1}(0, \infty; V')$. If we denote

$$\mathcal{P}(t) = \begin{pmatrix} \theta_\varepsilon(t) \\ G(t) \end{pmatrix}, \qquad (3.27)$$

where $G \in W^{1,\infty}(0, T; W^{1,1}(0, \infty; V'))$ is defined by

$$G(t)(s) := F(t + s), \ \forall s \in (0, \infty), \qquad (3.28)$$

we can write the problem

$$\frac{d\mathcal{P}}{dt}(t) + \mathcal{A}_\varepsilon(t)\mathcal{P}(t) = 0, \text{ a.e. } t \in (0, T), \qquad (3.29)$$

$$\mathcal{P}(0) = \begin{pmatrix} \theta_0 \\ F(s) \end{pmatrix}.$$

The second component of (3.29) provides the problem

$$\frac{\partial G(t, s)}{\partial t} - \frac{\partial G(t, s)}{\partial s} = 0,$$

$$G(0, s) = F(s),$$

that is necessarily satisfied by $G(t, s) = F(t + s)$. Then the problem corresponding to the first component in (3.29) reads

$$\frac{d\theta_\varepsilon}{dt}(t) + A_\varepsilon(t)\theta_\varepsilon(t) - G(t)(0) = 0, \quad \text{a.e. } t \in (0, T),$$

$$\theta_\varepsilon(0) = \theta_0.$$

Since $G(t)(0) = F(t)$ this turns out to be exactly (3.14).

The operator $A_\varepsilon(t)$ satisfies the same properties proved for $A_\varepsilon(t)$ in Lemma 3.2. Indeed, its domain is time-independent $D(A_\varepsilon(t)) = V \times W^{1,1}(0, \infty; V')$ and according to the arguments given in [46, 47] it turns out that the operator $A_\varepsilon(t)$ is quasi m-accretive on $\mathcal{X}$. Finally, the property (c) follows because

$$\|A_\varepsilon(t)\Theta - A_\varepsilon(\tau)\Theta\|_\mathcal{X}^2 = \|A_\varepsilon(t)\theta - A_\varepsilon(\tau)\theta\|_{V'}^2.$$

If $\theta_0 \in V$ it follows that $\frac{\theta_0}{u_0 + \varepsilon} \in V = D(A_\varepsilon(t))$ and consequently, we can apply the main result in [44] and assert that under the hypotheses (3.25) the problem (3.29) has a unique strong solution in the domain of the operator $A_\varepsilon(t)$, implying that problem (3.14) has a unique strong solution $\theta_\varepsilon \in C([0, T]; V') \cap W^{1,2}([0, T]; V')$ with $\beta_\varepsilon^*(\theta_\varepsilon) \in L^2(0, T; V)$. The latter implies also that $\theta_\varepsilon \in L^2(0, T; V)$. The solution satisfies the estimates

$$\int_\Omega u_\varepsilon(t, x) j_\varepsilon\left(\frac{\theta_\varepsilon}{u_\varepsilon}(t)\right) dx + \int_0^t \left\|\beta_\varepsilon^*\left(\frac{\theta_\varepsilon}{u_\varepsilon}(\tau)\right)\right\|_V^2 d\tau$$

$$+ \int_0^t \left\|\frac{d\theta_\varepsilon}{d\tau}(\tau)\right\|_{V'}^2 d\tau \le C \left(\int_\Omega (u_0 + \varepsilon) j_\varepsilon\left(\frac{\theta_0}{u_0 + \varepsilon}\right) dx\right.$$

$$\left. + \left\|\frac{\theta_0}{u_0}\right\|^2 + C_1 + 1\right), \quad \text{for } t \in [0, T], \tag{3.30}$$

$$\left\|\theta_\varepsilon(t) - \overline{\theta}_\varepsilon(t)\right\|_{V'} \le C(\varepsilon) C_2. \tag{3.31}$$

By C and $C(\varepsilon)$ we denote constants independent and dependent on ε, respectively, where

$$C_1 := 3\left\{\overline{K}^2 T + \int_0^T \left(\|f(t)\|_{V'}^2 + c_{tr}^2 \|p(t)\|_{L^2(\Gamma_p)}^2\right) dt\right\},$$

$$C_2 := \left\|\theta_0 - \overline{\theta}_0\right\|_{V'}^2 + \int_0^T \left(\|f(t) - \overline{f}(t)\|_{V'}^2 + c_{tr}^2 \|p(t) - \overline{p}(t)\|_{L^2(\Gamma_p)}^2\right) dt,$$

$$\overline{K} = K_s(\text{meas}(\Omega))^{1/2} \sum_{j=1}^N a_j^M,$$

the constant c_{tr} is given by the trace theorem and θ and $\overline{\theta}$ are the solutions corresponding to the pairs of data (θ_0, f, p) and $(\overline{\theta}_0, \overline{f}, \overline{p})$.

Let us notice that if $u \in W^{2,\infty}(Q)$ and $z \in V'$ we have $u_\varepsilon z \in V'$. Moreover, $(u_\varepsilon)_t \in W^{1,\infty}(Q)$ and if $\theta_\varepsilon \in L^2(0, T; V')$ then we have (in the sense of distributions) that

$$\frac{d\theta_\varepsilon}{dt} = u_\varepsilon \frac{d}{dt}\left(\frac{\theta_\varepsilon}{u_\varepsilon}\right) + \frac{\theta_\varepsilon}{u_\varepsilon}(u_\varepsilon)_t. \tag{3.32}$$

Also, we have the identity

$$\int_0^t \left\langle \left(u_\varepsilon \frac{d}{d\tau}\left(\frac{\theta_\varepsilon}{u_\varepsilon}\right)\right)(\tau), \beta_\varepsilon^*\left(\frac{\theta_\varepsilon}{u_\varepsilon}(\tau)\right)\right\rangle_{V',V} d\tau \tag{3.33}$$

$$= \int_\Omega u_\varepsilon(t, x) j_\varepsilon\left(\frac{\theta_\varepsilon}{u_\varepsilon}(t)\right) dx - \int_\Omega (u_0 + \varepsilon) j_\varepsilon\left(\frac{\theta_0}{u_0 + \varepsilon}\right) dx.$$

Indeed, let us define the function $\varphi_\varepsilon : (0, T) \times L^2(\Omega) \to (-\infty, \infty)$,

$$\varphi_\varepsilon(t, y) := \int_\Omega u_\varepsilon(t, x) j_\varepsilon(y) dx. \tag{3.34}$$

This function is differentiable with respect to y and t and

$$\partial_y \varphi_\varepsilon(t, y) = u_\varepsilon(t, x) \partial j_\varepsilon(y) = u_\varepsilon(t, x) \beta_\varepsilon^*(y).$$

Then we have

$$\int_0^t \left\langle \left(u_\varepsilon \frac{dy}{d\tau}\right)(\tau), \beta_\varepsilon^*(y(\tau))\right\rangle_{V',V} d\tau$$

$$= \int_0^t \left\langle \frac{dy}{d\tau}(\tau), u_\varepsilon(\tau) \beta_\varepsilon^*(y(\tau))\right\rangle_{V',V} d\tau$$

$$= \int_0^t \left\langle \frac{dy}{d\tau}(\tau), \partial_y \varphi_\varepsilon(\tau, \cdot)\right\rangle_{V',V} d\tau$$

$$= \int_0^t \frac{\partial \varphi_\varepsilon(\tau, y)}{\partial \tau} d\tau = \int_\Omega u_\varepsilon(t, x) j_\varepsilon(y) dx$$

$$- \int_\Omega (u_0 + \varepsilon) j_\varepsilon\left(\frac{\theta_0}{u_0 + \varepsilon}\right) dx.$$

This proves (3.33). Now, we test (3.14) at $\frac{\theta_\varepsilon}{u_\varepsilon}$ and integrate over $(0, t)$. We obtain

$$\int_0^t \left\langle \frac{d\theta_\varepsilon}{d\tau}(\tau), \frac{\theta_\varepsilon}{u_\varepsilon}(\tau) \right\rangle_{V',V} d\tau + \int_0^t \int_\Omega \nabla \beta_\varepsilon^* \left(\frac{\theta_\varepsilon}{u_\varepsilon} \right) \cdot \nabla \left(\frac{\theta_\varepsilon}{u_\varepsilon} \right) dx d\tau$$

$$\leq \int_0^t \int_\Omega K_0 \left(\tau, x, \frac{\theta_\varepsilon}{u_\varepsilon} \right) \cdot \nabla \left(\frac{\theta_\varepsilon}{u_\varepsilon} \right) dx d\tau$$

$$+ \int_0^t \|f(\tau)\|_{V'} \left\| \frac{\theta_\varepsilon}{u_\varepsilon}(\tau) \right\|_V d\tau + \int_0^t \|p(\tau)\|_{L^2(\Gamma_p)} \left\| \frac{\theta_\varepsilon}{u_\varepsilon}(\tau) \right\|_{L^2(\Gamma_p)} d\tau.$$

Since $\sqrt{u_\varepsilon} \in W^{2,\infty}(Q)$ and $u_\varepsilon \geq \varepsilon > 0$ we have $\frac{d\sqrt{u_\varepsilon}}{dt} = \frac{(u_\varepsilon)_t}{2\sqrt{u_\varepsilon}} \in W^{1,\infty}(Q)$.
We write $\theta_\varepsilon = \frac{\theta_\varepsilon}{\sqrt{u_\varepsilon}}\sqrt{u_\varepsilon}$ and apply (3.32),

$$\int_0^t \left\langle \sqrt{u_\varepsilon}(\tau) \frac{d}{d\tau} \left(\frac{\theta_\varepsilon}{\sqrt{u_\varepsilon}} \right)(\tau), \frac{\theta_\varepsilon}{u_\varepsilon}(\tau) \right\rangle_{V',V} d\tau$$

$$+ \int_0^t \int_\Omega \frac{\theta_\varepsilon}{\sqrt{u_\varepsilon}} \frac{(u_\varepsilon)_t}{2\sqrt{u_\varepsilon}} \frac{\theta_\varepsilon}{u_\varepsilon} dx d\tau + \rho \int_0^t \left\| \frac{\theta_\varepsilon}{u_\varepsilon}(\tau) \right\|_V^2 d\tau$$

$$\leq \int_0^t \left\| K_0 \left(\tau, x, \frac{\theta_\varepsilon}{u_\varepsilon}(\tau) \right) \right\| \left\| \frac{\theta_\varepsilon}{u_\varepsilon}(\tau) \right\|_V d\tau$$

$$+ \int_0^t \|f(\tau)\|_{V'} \left\| \frac{\theta_\varepsilon}{u_\varepsilon}(\tau) \right\|_V d\tau$$

$$+ c_{tr} \int_0^t \|p(\tau)\|_{L^2(\Gamma_p)} \left\| \frac{\theta_\varepsilon}{u_\varepsilon}(\tau) \right\|_V d\tau,$$

where c_{tr} is the constant in the trace theorem. After a few calculations we get

$$\int_0^t \left\langle \frac{d}{d\tau} \left(\frac{\theta_\varepsilon}{\sqrt{u_\varepsilon}} \right)(\tau), \frac{\theta_\varepsilon}{\sqrt{u_\varepsilon}}(\tau) \right\rangle_{V',V} d\tau + \rho \int_0^t \left\| \frac{\theta_\varepsilon}{u_\varepsilon}(\tau) \right\|_V^2 d\tau$$

$$\leq \frac{\rho}{2} \int_0^t \left\| \frac{\theta_\varepsilon}{u_\varepsilon}(\tau) \right\|_V^2 d\tau + \frac{C_1}{2\rho} + \frac{1}{2} \int_0^t \int_\Omega \left(\frac{\theta_\varepsilon}{u_\varepsilon} \right)^2 |(u_\varepsilon)_\tau| d\tau dx.$$

Integrating the first term on the left-hand side with respect to τ we obtain

$$\left\| \frac{\theta_\varepsilon}{\sqrt{u_\varepsilon}}(t) \right\|^2 + \rho \int_0^t \left\| \frac{\theta_\varepsilon}{u_\varepsilon}(\tau) \right\|_V^2 d\tau \leq \left\| \frac{\theta_0}{\sqrt{u_0 + \varepsilon}} \right\|^2$$

$$+ \frac{C_1}{\rho} + c_P^2 u_M' \int_0^t \left\| \frac{\theta_\varepsilon}{u_\varepsilon}(\tau) \right\|_V^2 d\tau,$$

hence,

$$\left\|\frac{\theta_\varepsilon}{\sqrt{u_\varepsilon}}(t)\right\|^2 + \rho_1 \int_0^t \left\|\frac{\theta_\varepsilon}{u_\varepsilon}(\tau)\right\|_V^2 d\tau \le (u_M + 1)\left\|\frac{\theta_0}{u_0}\right\|^2 + \frac{C_1}{\rho}, \qquad (3.35)$$

with $\rho_1 = \rho - c_P^2 u_M' > 0$, by (3.23). Indeed, since $u_0 \le u_0 + \varepsilon$ we have that

$$\left\|\frac{\theta_0}{\sqrt{u_0 + \varepsilon}}\right\| \le \sqrt{u_M + \varepsilon}\left\|\frac{\theta_0}{u_0}\right\|. \qquad (3.36)$$

To prove estimate (3.30) we multiply (3.14) by $\beta_\varepsilon^*\left(\frac{\theta_\varepsilon}{u_\varepsilon}\right)$ and integrate over $(0, t)$, writing $\theta_\varepsilon = u_\varepsilon \frac{\theta_\varepsilon}{u_\varepsilon}$. Using (3.32) and (3.33) we obtain after some computations that

$$\int_\Omega u_\varepsilon j_\varepsilon\left(\frac{\theta_\varepsilon}{u_\varepsilon}(t)\right) dx + \int_0^t \int_\Omega \frac{\theta_\varepsilon}{u_\varepsilon}\beta_\varepsilon^*\left(\frac{\theta_\varepsilon}{u_\varepsilon}\right)(u_\varepsilon)_\tau dx d\tau$$

$$+\frac{1}{2}\int_0^t \left\|\beta_\varepsilon^*\left(\frac{\theta_\varepsilon}{u_\varepsilon}(\tau)\right)\right\|_V^2 d\tau$$

$$\le \int_\Omega (u_0 + \varepsilon) j_\varepsilon\left(\frac{\theta_0}{u_0 + \varepsilon}\right) dx + \frac{C_1}{2}.$$

Further, writing that

$$\left|\int_0^t \int_\Omega \frac{\theta_\varepsilon}{u_\varepsilon}\beta_\varepsilon^*\left(\frac{\theta_\varepsilon}{u_\varepsilon}\right)(u_\varepsilon)_\tau dx d\tau\right|$$

$$\le \int_0^t \left\|\beta_\varepsilon^*\left(\frac{\theta_\varepsilon}{u_\varepsilon}(\tau)\right)\right\|\left\|\frac{\theta_\varepsilon}{u_\varepsilon}(\tau)(u_\varepsilon)_\tau(\tau)\right\| d\tau$$

$$\le c_P \int_0^t \left\|\beta_\varepsilon^*\left(\frac{\theta_\varepsilon}{u_\varepsilon}(\tau)\right)\right\|_V \left\|\frac{\theta_\varepsilon}{u_\varepsilon}(\tau)(u_\varepsilon)_\tau(\tau)\right\| d\tau$$

$$\le \frac{1}{4}\int_0^t \left\|\beta_\varepsilon^*\left(\frac{\theta_\varepsilon}{u_\varepsilon}(\tau)\right)\right\|_V^2 d\tau + c_P^2 \int_0^t \left\|\frac{\theta_\varepsilon}{u_\varepsilon}(\tau)(u_\varepsilon)_\tau^2(\tau)\right\| d\tau$$

we have

$$\int_\Omega u_\varepsilon j_\varepsilon\left(\frac{\theta_\varepsilon}{u_\varepsilon}(t)\right) dx + \frac{1}{2}\int_0^t \left\|\beta_\varepsilon^*\left(\frac{\theta_\varepsilon}{u_\varepsilon}(\tau)\right)\right\|_V^2 d\tau$$

$$\le \int_\Omega (u_0 + \varepsilon) j_\varepsilon\left(\frac{\theta_0}{u_0 + \varepsilon}\right) dx + C_1 + \frac{1}{4}\int_0^t \left\|\beta_\varepsilon^*\left(\frac{\theta_\varepsilon}{u_\varepsilon}(\tau)\right)\right\|_V d\tau$$

$$+c_P^2 \int_0^t \int_\Omega \left(\frac{\theta_\varepsilon}{u_\varepsilon}\right)^2 (u_\varepsilon)_\tau^2 dx d\tau.$$

Since $(u_\varepsilon)_\tau \leq u'_M$ we get

$$
\int_\Omega u_\varepsilon(t,x) j_\varepsilon \left(\frac{\theta_\varepsilon}{u_\varepsilon}(t) \right) dx + \int_0^t \left\| \beta_\varepsilon^* \left(\frac{\theta_\varepsilon}{u_\varepsilon}(\tau) \right) \right\|_V^2 d\tau
$$
$$
\leq 4 \left(\int_\Omega (u_0 + \varepsilon) j_\varepsilon \left(\frac{\theta_0}{u_0 + \varepsilon} \right) dx + C_1 \right)
$$
$$
+ 4 c_P^4 (u'_M)^2 \int_0^t \left\| \frac{\theta_\varepsilon}{u_\varepsilon}(t) \right\|_V^2 d\tau.
$$

Using (3.35) we obtain

$$
\int_\Omega u_\varepsilon(t,x) j_\varepsilon \left(\frac{\theta_\varepsilon}{u_\varepsilon}(t) \right) dx + \int_0^t \left\| \beta_\varepsilon^* \left(\frac{\theta_\varepsilon}{u_\varepsilon}(\tau) \right) \right\|_V^2 d\tau \tag{3.37}
$$
$$
\leq C \left(\int_\Omega (u_0 + \varepsilon) j_\varepsilon \left(\frac{\theta_0}{u_0 + \varepsilon} \right) dx + \left\| \frac{\theta_0}{u_0} \right\|^2 + C_1 + 1 \right)
$$

with C depending on the problem data and independent on ε.

We multiply now (3.14) by $\frac{d\theta_\varepsilon}{dt}(t)$ in V' and integrate over $(0,t)$. We have

$$
\int_0^t \left\| \frac{d\theta_\varepsilon}{d\tau}(\tau) \right\|_{V'}^2 d\tau + \int_0^t \left\langle \frac{d\theta_\varepsilon}{d\tau}(\tau), \beta_\varepsilon^* \left(\frac{\theta_\varepsilon}{u_\varepsilon}(\tau) \right) \right\rangle_{V',V} d\tau
$$
$$
= \int_0^t \int_\Omega K_0 \left(\tau, x, \frac{\theta_\varepsilon}{u_\varepsilon} \right) \cdot \nabla \beta_\varepsilon^* \left(\frac{\theta_\varepsilon}{u_\varepsilon} \right) dx d\tau
$$
$$
+ \int_0^t \left\langle f(\tau), \beta_\varepsilon^* \left(\frac{\theta_\varepsilon}{u_\varepsilon} \right) \right\rangle_{V',V} d\tau - \int_0^t \int_{\Gamma_p} p \beta_\varepsilon^* \left(\frac{\theta_\varepsilon}{u_\varepsilon} \right) dx d\tau.
$$

Since the computations are similar to those made before we do not present them in detail. We obtain

$$
\int_\Omega u_\varepsilon(t,x) j_\varepsilon \left(\frac{\theta_\varepsilon}{u_\varepsilon}(t) \right) dx + \int_0^t \left\| \frac{d\theta_\varepsilon}{d\tau}(\tau) \right\|_{V'}^2 d\tau \tag{3.38}
$$
$$
\leq C \left(\int_\Omega (u_0 + \varepsilon) j_\varepsilon \left(\frac{\theta_0}{u_0 + \varepsilon} \right) dx + \left\| \frac{\theta_0}{u_0} \right\|^2 + C_1 + 1 \right).
$$

Summing up the last two relations we get (3.30) as claimed.

Relation (3.31) is proved exactly like (1.53).

Step 2. In the second step we assume

$$f, \ f_{\Gamma_p} \in L^2(0, T; V'), \ \theta_0 \in L^2(\Omega).$$

We can extend f and f_{Γ_p} by 0 for $t \in (T, +\infty)$ and let us consider the sequences

$$(f_n)_{n \geq 1} \in W^{1,1}(0, \infty; V'), \ (f_{\Gamma_p}^n)_{n \geq 1} \in W^{1,1}(0, \infty; V'), \ (\theta_0^n)_{n \geq 1} \subset V$$

such that

$$F_n = f_n + f_{\Gamma_p}^n \to F = f + f_{\Gamma_p} \text{ in } L^2(0, T; V') \text{ as } n \to \infty$$

and

$$\theta_0^n \to \theta_0 \text{ in } L^2(\Omega), \text{ as } n \to \infty.$$

Then we consider the problem

$$\frac{d\mathcal{P}_n}{dt}(t) + \mathcal{A}_\varepsilon(t)\mathcal{P}_n(t) = 0, \text{ a.e. } t \in (0, T), \tag{3.39}$$

$$\mathcal{P}_n(0) = \begin{pmatrix} \theta_0^n \\ F_n(s) \end{pmatrix}.$$

For the second component (3.28) we have

$$G_n(t)(s) = F_n(t + s) \to F(t + s) = G(t)(s), \text{ as } n \to \infty, \ \forall s \in (0, \infty),$$

while the first component reads

$$\frac{d\theta_\varepsilon^n}{dt}(t) + A_\varepsilon(t)\theta_\varepsilon^n(t) = F_n(t), \text{ a.e. } t \in (0, T), \tag{3.40}$$

$$\theta_\varepsilon^n(0) = \theta_0^n.$$

By the previous step we know that (3.39) has a unique solution, meaning that (3.40) has a solution satisfying (3.30). From here, by passing to limit as $n \to \infty$, according to the second step in Proposition 1.5 we obtain that (3.14) has a solution which is unique on the basis of (3.31). Consequently (3.29) has a unique solution. Anyway, further we are interested only in (3.14).

Step 3. In this part we pass to the limit as $\varepsilon \to 0$.
Let us assume (3.4) and $f, \ f_{\Gamma_p} \in L^2(0, T; V')$. According to the second step (3.14) has a unique solution satisfying (3.30). We can write

$$\int_\Omega j_\varepsilon\left(\frac{\theta_0}{u_0+\varepsilon}\right)dx = \int_\Omega \int_0^{\theta_0/(u_0+\varepsilon)} \beta_\varepsilon^*(r)dr\,dx$$

$$\leq \int_{\Omega_u} \int_0^{\theta_0/u_0} \beta_\varepsilon^*(r)dr\,dx$$

$$\leq \int_{\Omega_u} \int_0^{y_s} \beta_\varepsilon^*(r)dr \leq \beta_s^* y_s \operatorname{meas}(\Omega).$$

Plugging this result in (3.30) we obtain

$$\int_\Omega u_\varepsilon(t,x)j_\varepsilon\left(\frac{\theta_\varepsilon}{u_\varepsilon}(t)\right)dx + \int_0^t \left\|\beta_\varepsilon^*\left(\frac{\theta_\varepsilon}{u_\varepsilon}(\tau)\right)\right\|_V^2 d\tau$$

$$+ \int_0^t \left\|\frac{d\theta_\varepsilon}{d\tau}(\tau)\right\|_V^2 d\tau \leq C, \tag{3.41}$$

for any $t \in [0,T]$, where C does not depend on ε. The estimates (3.35) and (3.41) allow us to deduce that by selecting successive subsequences we get

$$\beta_\varepsilon^*\left(\frac{\theta_\varepsilon}{u_\varepsilon}\right) \rightharpoonup \zeta \text{ in } L^2(0,T;V), \text{ as } \varepsilon \to 0, \tag{3.42}$$

$$\frac{\theta_\varepsilon}{u_\varepsilon} \rightharpoonup y \text{ in } L^2(0,T;V), \text{ as } \varepsilon \to 0, \tag{3.43}$$

$$\theta_\varepsilon \rightharpoonup \theta \text{ in } L^2(0,T;V), \text{ as } \varepsilon \to 0, \tag{3.44}$$

$$\frac{d\theta_\varepsilon}{dt} \rightharpoonup \frac{d\theta}{dt} \text{ in } L^2(0,T;V'), \text{ as } \varepsilon \to 0. \tag{3.45}$$

Following the arguments presented in Theorem 1.6 we deduce

$$\theta = uy, \quad \zeta \in \beta^*(y) \text{ a.e. on } Q, \tag{3.46}$$

and

$$K_0\left(t,x,\frac{\theta_\varepsilon}{u_\varepsilon}\right) \rightharpoonup K_0(t,x,y) \text{ in } L^2(0,T;L^2(\Omega)), \text{ as } \varepsilon \to 0. \tag{3.47}$$

Finally we can pass to the limit as $\varepsilon \to 0$ in (3.16) and get

$$\int_0^T \left\langle \frac{d(uy)}{dt}(t), \phi(t) \right\rangle_{V',V} dt$$

$$+ \int_Q \left(\nabla \zeta \cdot \nabla \phi - K_0\left(t, x, y\right) \cdot \nabla \phi \right) dx dt$$

$$= \int_0^T \langle f(t), \phi(t) \rangle_{V',V} \, dt - \int_{\Sigma_p} p \phi d\sigma dt, \ \ \forall \phi \in L^2(0,T;V),$$

for any $\phi \in L^2(0,T;V)$. Next we proceed as for the solution construction in Theorem 1.6 and for uniqueness as in Proposition 1.8. $\square$

Another possibility of treating nonautonomous problems is by following the Brezis–Ekeland principle, see [31, 32]. This approach based on the Legendre–Fenchel–Young relations allows the study of equations with less regular data and leads to generalized solutions in larger spaces of functions. Such results have been obtained in [90] for (3.1) in the fast diffusion case and in the case with β and β^* defined on $\mathbb{R}$, i.e., case (c) from Introduction.

3.3.1 Numerical Results

The numerical results in this section are presented for the same data (Ω, Ω_0, β^*, K_0, f, θ_0) as in Sect. 1.1 except for the porosity u which is considered increasing in time

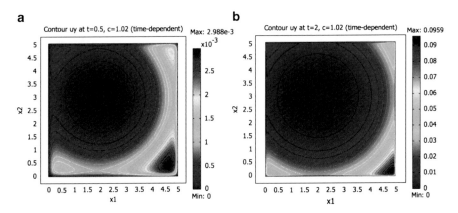

Fig. 3.1 Solution $\theta = uy$ in the parabolic–elliptic degenerate case for the time-dependent u given by (3.48) and $c = 1.02$

$$u(t, x_1, x_2) := \begin{cases} 0, & \text{on } Q_0 \\ \left(\frac{(x_1-2)^2 + (x_2-3)^2 - 0.1^2}{100} \right)^2 t^2, & \text{on } Q_u. \end{cases} \tag{3.48}$$

The graphics of $\theta = uy$ for $c = 1.02$ are shown in Fig. 3.1a, b at $t = 0.5$, 2. In this case the porosity is much smaller than in Sect. 1.1. We can notice the influence of the time variable porosity upon the solution by the increase of the water content values at larger times $(t = 2)$ as u increases.

Chapter 4
Parameter Identification in a Parabolic–Elliptic Degenerate Problem

In applied sciences an important achievement is the determination of the parameters of the equations modelling a physical process from the knowledge of certain physical quantities which can be observed or measured. These are formulated as identification and inverse problems that can be treated as optimal control problems. This chapter is devoted to such an example.

We shall study an identification problem in relation with (1.1). More specifically, the purpose is to identify the space variable time derivative coefficient u from available observations on the function $\theta = uy$. This problem can be approached as a control problem (P) with the control u in $W^{2,m}(\Omega)$ for $m > N$. First, we introduce an approximating control problem (P_ε) and prove the existence of an optimal pair. Under certain assumptions on the data the necessary conditions that should be obeyed by the control are found in an implicit variational form. Next, it is shown that a sequence of optimal pairs $(u_\varepsilon^*, y_\varepsilon^*)_{\varepsilon>0}$ for (P_ε) converges as ε goes to 0 to a pair (u^*, y^*) which realizes the minimum in (P), and y^* is a solution to the original state system. An alternative way of handling the control problem can be led by considering two controls related between them by a certain elliptic problem. This approach enables the determination of simpler conditions of optimality under an additional restriction upon the initial datum of the direct problem. We also remark that inverse problems are ill-posed and in general we do not have uniqueness.

The problem without the advection term $(a(x) = 0)$ was treated in [60].

4.1 Statement of the Problem

Let us resume (1.1), i.e., the parabolic–elliptic degenerate problem

$$\frac{\partial}{\partial t}(u(x)y) - \Delta\beta^*(y) + \nabla \cdot (a(x)K(y)) \ni f \text{ in } Q, \qquad (4.1)$$

A. Favini and G. Marinoschi, *Degenerate Nonlinear Diffusion Equations,*
Lecture Notes in Mathematics 2049, DOI 10.1007/978-3-642-28285-0_4,
© Springer-Verlag Berlin Heidelberg 2012

$$y(t, x) = 0 \text{ on } \Sigma, \tag{4.2}$$

$$(u(x)y(t, x))|_{t=0} = \theta_0 \text{ in } \Omega. \tag{4.3}$$

In this chapter the dimension of the space is taken $N \leq 3$ and we consider $\partial \Omega$ of class C^2. The parameters satisfy all conditions (1.2)–(1.14) and in addition we require that $\beta \in C^2(-\infty, y_s)$. We recall that

$$0 \leq u \leq u_M, \quad u = 0 \text{ on } \overline{\Omega_0}, \quad u > 0 \text{ on } \Omega_u = \Omega \backslash \overline{\Omega_0}, \tag{4.4}$$

and

$$\Omega_0 = \{x \in \Omega; u(x) = 0\}, \quad \Omega_u = \{x \in \Omega; u(x) > 0\}.$$

The identification problem will be approached under stronger assumptions on the problem data, respectively,

$$f \in W^{1,2}([0, T]; L^2(\Omega)), \tag{4.5}$$

$$\theta_0 \in H^2(\Omega) \cap H_0^1(\Omega), \quad \theta_0 = 0 \text{ in } \Omega_0, \tag{4.6}$$

$$\theta_0 \geq 0 \text{ in } \Omega_u, \quad \frac{\theta_0}{u} \in L^2(\Omega_u), \quad \frac{\theta_0}{u} \leq y_s, \text{ for } x \in \Omega_u,$$

and u will be found in a more restrictive class of functions, $W^{2,m}(\Omega)$, with $m > N$.

We introduce the function $\theta(t, x) = u(x)y(t, x)$, and assume that there are available data $\theta_{obs}(t, x)$ observed for θ in a domain $Q_{obs} = (0, T_{obs}) \times \Omega_{obs}$, where $T_{obs} \leq T$ and $\Omega_{obs} \subseteq \Omega$, $\theta_{obs} \in L^2(Q_{obs})$. Denoting

$$J(u) = \frac{1}{2} \int_{Q_{obs}} (u(x)y(t, x) - \theta_{obs}(t, x))^2 dx dt + \frac{k_1}{m} \int_{\Omega} (u(x) - \Delta u(x))^m dx$$

we are going to approach the identification problem as an optimal control problem, by searching for the control u as the solution to the problem

$$\text{Minimize} J(u) \tag{P}$$

subject to (4.1)–(4.3), for all $u \in U$, where

$$U = \left\{ u \in W^{2,m}(\Omega), \frac{\theta_0}{y_s}(x) \leq u(x) \leq u_M \text{ on } \Omega, \left. \frac{\partial u}{\partial \nu} \right|_{\Gamma} = 0, u = 0 \text{ on } \overline{\Omega_0} \right\}, \tag{4.7}$$

and m is a positive even integer,

$$m = 2k > N, \quad k \geq 1. \tag{4.8}$$

In a rigorous way, in (P) we should have denoted y^u in order to specify that y depends on u, but for not complicating the notation we make the convention to skip the superscript indicating u. The positive constant k_1 is a weight that may be chosen in such a way to enhance a greater importance to a term in (P) against the other.

The last term on the right-hand side in (P) was introduced to induce the regularity $u \in W^{2,m}(\Omega)$ necessary in the next proofs. On the one hand this implies by the Sobolev embedding theorem that $W^{2,m}(\Omega)$ is compactly embedded in $W^{1,\infty}(\Omega)$ for $m > N$. Also u belongs to $H^2(\Omega)$, so that the control $u \in H^2(\Omega) \cap W^{1,\infty}(\Omega)$. The property $u \in W^{1,\infty}(\Omega)$ is necessary to ensure that u is a multiplier in V' while $u \in H^2(\Omega)$ will be used in the proof of existence of the solution to the dual system. For $N \leq 2$ it is sufficient to take $m = 2$.

We make a notation that will be further used. For $u \in W^{2,m}(\Omega)$ we denote

$$v := u - \Delta u \in L^m(\Omega).$$

We also recall that according to the result due to Agmon et al. (see [1]), the problem

$$u - \Delta u = v \text{ in } \Omega, \tag{4.9}$$

$$\nabla u \cdot \nu = 0 \text{ on } \Gamma$$

with $v \in L^m(\Omega)$, has a unique solution $u \in W^{2,m}(\Omega)$, such that

$$\|u\|_{W^{2,m}(\Omega)} \leq C \|u\|_{L^m(\Omega)} \text{ for any } u \in L^m(\Omega). \tag{4.10}$$

The inequality $\frac{\theta_0}{y_s} \leq u$ is required in the proof of existence in the state system, while $u \leq u_M$ is mostly related to the physical interpretation of u.

We mention that from the mathematical point of view, Robin or nonhomogeneous Dirichlet boundary conditions might also be considered in U instead of the Neumann one. A homogeneous Dirichlet condition should be avoided because $\overline{\Omega_0}$ was considered to be strictly included in Ω.

Because the state system involves a multivalued operator and in the perspective of establishing a framework for numerical computations an appropriate approximating problem (P_ε) indexed on a small positive parameter ε will be introduced. It involves an approximating state system. The approach of the identification problem will address the following aspects: existence, uniqueness and regularity of the approximating state system; proof of the existence of a solution to (P_ε); computation of the approximate optimality condition, after introducing and studying the system of first order variations and the dual system for the approximating problem; proof of the fact that (P_ε) approximates (P), i.e., that a sequence of optimal pairs $(u_\varepsilon^*, y_\varepsilon^*)_\varepsilon$ for (P_ε) tends to a pair (u^*, y^*) which realizes the minimum in (P) and the state y^* corresponding to u^* is the solution to (4.1)–(4.3).

4.2   The Approximating Control Problem

Let ε be positive, small and consider a smooth approximation of the multi-valued function β^*. For the purposes of this part β^* is replaced by a three times differentiable function $\beta_\varepsilon^* : \mathbb{R} \to \mathbb{R}$, taken for example as

$$\beta_\varepsilon^*(r) = \begin{cases} \beta^*(r), & r < y_s - \varepsilon \\ \beta_{reg}^*(r), & y_s - \varepsilon \le r \le y_s \\ \beta^*(y_s - \varepsilon) + \frac{\beta_s^* - \beta^*(y_s - \varepsilon)}{\varepsilon}[r - (y_s - \varepsilon)], & r > y_s, \end{cases} \quad (4.11)$$

where β_{reg}^* is a regular function that can be simply constructed (taking into account the properties (1.4)–(1.6)), such that β_ε^* and its first three derivatives denoted by β_ε, β_ε', β_ε'', respectively, have the properties

$$\lim_{r \to \pm\infty} \beta_\varepsilon^*(r) = \pm\infty,$$

$$\beta_\varepsilon(r) \ge \rho > 0 \text{ for } r \in \mathbb{R}, \quad \beta_\varepsilon(r) = \beta(r) = \rho \text{ for } r \le 0, \quad (4.12)$$

$$\beta_\varepsilon'(r) = 0 \text{ for } r \in (-\infty, 0] \cup [y_s, \infty), \ \beta_\varepsilon'' \in L^\infty(\mathbb{R}).$$

The function K is extended at the right of $r = y_s$ by $K(y_s)$. Let us denote

$$J_\varepsilon(u) = \frac{1}{2} \int_{Q_{obs}} ((u + \varepsilon)y(t, x) - \theta_{obs}(t, x))^2 dx dt \quad (4.13)$$

$$+ \frac{k_1}{m} \int_\Omega v^m(x) dx + \frac{1}{2\varepsilon} \int_{\Omega_0} u^2(x) dx,$$

$$\tilde{U} = \left\{ u \in W^{2,m}(\Omega); \ \frac{\theta_0}{y_s}(x) \le u(x) \le u_M \text{ on } \Omega, \ \frac{\partial u}{\partial \nu}\bigg|_\Gamma = 0 \right\},$$

$$(4.14)$$

where $v = u - \Delta u$, and introduce the problem

$$\text{Minimize } J_\varepsilon(u), \qquad (P_\varepsilon)$$

for all $u \in \tilde{U}$, subject to the approximating problem

$$\frac{\partial}{\partial t}((u + \varepsilon)y) - \Delta\beta_\varepsilon^*(y) + \nabla \cdot (a(x)K(y)) = f \text{ in } Q, \quad (4.15)$$

$$y(t, x) = 0 \text{ on } \Sigma, \quad (4.16)$$

$$(u + \varepsilon)y(0, x) = \theta_0 \text{ in } \Omega. \quad (4.17)$$

We notice that $U \subset \tilde{U}$. The last term introduced on the right-hand side in (4.13) will force the vanishing of u on $\overline{\Omega_0}$ as ε will go to zero.

4.2.1 Existence in the Approximating State System

We write the abstract Cauchy problem (equivalent with (4.15)–(4.17))

$$\frac{d((u+\varepsilon)y_\varepsilon)}{dt}(t) + A_\varepsilon y_\varepsilon(t) = f(t) \text{ a.e. } t \in (0,T), \tag{4.18}$$

$$u_\varepsilon y_\varepsilon(0) = \theta_0,$$

with $u \in W^{2,m}(\Omega)$, $u \geq 0$ and denote $\theta_\varepsilon(t,x) := (u+\varepsilon)y_\varepsilon(t,x)$. The operator A_ε is that defined in (1.38). As specified before we begin by proving the results for the approximating problem, in the same functional space as in Sect. 1.1. We recall Proposition 1.5 and add also a regularity result.

Proposition 4.1. *Assume (4.5), (4.6), and*

$$u \in W^{2,m}(\Omega), \quad 0 \leq u \leq u_M. \tag{4.19}$$

Then, the Cauchy problem (4.18) has a unique solution

$$y_\varepsilon, \theta_\varepsilon, \beta_\varepsilon^*(y_\varepsilon) \in L^2(0,T;V) \cap W^{1,2}([0,T];V'), \tag{4.20}$$

satisfying

$$\int_\Omega (u+\varepsilon) j_\varepsilon \left(\frac{\theta_\varepsilon}{u+\varepsilon}(t) \right) dx + \frac{1}{4} \int_0^t \left\| \frac{d\theta_\varepsilon}{d\tau}(\tau) \right\|_{V'}^2 d\tau$$

$$+ \frac{1}{4} \int_0^t \left\| \beta_\varepsilon^* \left(\frac{\theta_\varepsilon}{u}(\tau) \right) \right\|_V^2 d\tau$$

$$\leq \int_\Omega (u+\varepsilon) j_\varepsilon \left(\frac{\theta_0}{u+\varepsilon} \right) dx + \int_0^T \|f(t)\|_{V'}^2 dt + \overline{K}^2 T, \ t \in [0,T]. \tag{4.21}$$

In addition, we have

$$y_\varepsilon, \theta_\varepsilon, \beta_\varepsilon^*(y_\varepsilon) \in W^{1,\infty}([0,T];L^2(\Omega)) \cap W^{1,2}([0,T];V) \cap L^\infty(0,T;H^2(\Omega)), \tag{4.22}$$

$$\|\beta_\varepsilon^*(y_\varepsilon)\|_{W^{1,\infty}([0,T];L^2(\Omega))}^2 + \|\beta_\varepsilon^*(y_\varepsilon)\|_{W^{1,2}([0,T];V)}^2$$

$$+ \|\beta_\varepsilon^*(y_\varepsilon)\|_{L^\infty(0,T;H^2(\Omega))}^2 \leq \gamma_0 C(\varepsilon), \tag{4.23}$$

$$\|y_\varepsilon\|_{W^{1,\infty}([0,T];L^2(\Omega))}^2 + \|y_\varepsilon\|_{W^{1,2}([0,T];V)}^2 + \|y_\varepsilon\|_{L^\infty(0,T;H^2(\Omega))}^2 \leq \gamma_0 C(\varepsilon), \tag{4.24}$$

where γ_0 denotes several constants dependent on N, Ω, ρ, T and

$$C(\varepsilon) = \gamma_0 \beta_M(\varepsilon) \times \left(\left\| \beta_\varepsilon^* \left(\frac{\theta_0}{u + \varepsilon} \right) \right\|_V^2 \right.$$

$$\left. + \int_\Omega j_\varepsilon \left(\frac{\theta_0}{u + \varepsilon} \right) dx + \int_0^T \| f(t) \|_{L^2(\Omega)}^2 \, dt + 1 \right),$$

$$\beta_M(\varepsilon) = \max_{r \in \mathbb{R}} \beta_\varepsilon(r) = \frac{\beta_s^* - \beta^*(y_s - \varepsilon)}{\varepsilon}. \tag{4.25}$$

Proof. By (4.6) and (4.19) it follows by a straightforward computation that

$$\frac{\theta_0}{u + \varepsilon} \in H_0^1(\Omega) \cap H^2(\Omega). \tag{4.26}$$

Then, Proposition 1.5 gives the first part of the solution existence (4.20), its uniqueness and estimate (4.21). The other estimates (4.22)–(4.24) are proved by a very technical proof. A similar proof was done in [83] (see also [84], Theorem 2.8, pp. 160) for a 3D model with an advection term $a(x) = (0, 0, 1)$ and Robin boundary conditions. With slight modifications corresponding to the Dirichlet boundary conditions the argument applies here as well, so that we shall not develop the computations and refer the reader to these works. We mention only that the hypotheses under which this result takes place, i.e., (4.5) and (4.26) are satisfied. Finally, since $u_\varepsilon \in W^{2,m}(\Omega)$ it follows that θ_ε belongs to the same spaces as y_ε and satisfies (4.24).   □

4.2.2 Existence of the Approximating Optimal Control

Theorem 4.2. *Assume (4.5) and (4.6). Then, problem (P_ε) has at least one solution $u_\varepsilon^* \in \tilde{U}$. The corresponding state y_ε^*, which is the solution to (4.18) with $u = u_\varepsilon^*$, satisfies (4.21)–(4.24).*

Proof. Let ε be fixed. Since $J_\varepsilon(u) \geq 0$, it follows that $d_\varepsilon = \inf_{u \in \tilde{U}} J_\varepsilon(u)$ exists and it is nonnegative. Let $(u_\varepsilon^n)_{n \geq 1}$ be a minimizing sequence with $u_\varepsilon^n \in \tilde{U}$. Then

$$d_\varepsilon \leq \frac{1}{2} \int_{Q_{obs}} ((u_\varepsilon^n + \varepsilon) y_\varepsilon^n(t, x) - \theta_{obs}(t, x))^2 dx dt \tag{4.27}$$

$$+ \frac{k_1}{m} \int_\Omega (v_\varepsilon^n)^m(x) dx + \frac{1}{2\varepsilon} \int_{\Omega_0} (u_\varepsilon^n)^2(x) dx \leq d_\varepsilon + \frac{1}{n},$$

where

$$v_\varepsilon^n = u_\varepsilon^n - \Delta u_\varepsilon^n \tag{4.28}$$

and y_ε^n is the solution to (4.18) corresponding to u_ε^n, i.e.,

$$\frac{d((u_\varepsilon^n + \varepsilon)y_\varepsilon^n)}{dt}(t) + A_\varepsilon y_\varepsilon^n(t) = f(t) \text{ a.e. } t \in (0, T), \tag{4.29}$$

$$u_\varepsilon^n y_\varepsilon^n(0) = \theta_0.$$

This is written in the equivalent form

$$\int_Q \frac{d((u_\varepsilon^n + \varepsilon)y_\varepsilon^n)}{dt} \psi dx dt + \int_Q (\nabla \beta_\varepsilon^*(y_\varepsilon^n) - a(x)K(y_\varepsilon^n)) \cdot \nabla \psi dx dt = \int_Q f\psi dx dt, \tag{4.30}$$

for any $\psi \in L^2(0, T; V)$,

$$\theta_\varepsilon^n(0) = (u_\varepsilon^n + \varepsilon)y_\varepsilon^n(0) = \theta_0, \tag{4.31}$$

where $\theta_\varepsilon^n = (u_\varepsilon^n + \varepsilon)y_\varepsilon^n$. By (4.27) we get on a subsequence, still denoted by the subscript n, that

$$v_\varepsilon^n \rightharpoonup v_\varepsilon^* \text{ in } L^m(\Omega) \text{ as } n \to \infty. \tag{4.32}$$

Now, $u_\varepsilon^n \in \widetilde{U}$, hence $\frac{\partial u_\varepsilon^n}{\partial \nu} = 0$ on Γ. Coupling it with (4.28) it follows that u_ε^n satisfies (4.9) with $v = v_\varepsilon^n$. Consequently, $\|u_\varepsilon^n\|_{W^{2,m}(\Omega)} \leq C \|v_\varepsilon^n\|_{L^m(\Omega)}$, hence

$$u_\varepsilon^n \rightharpoonup u_\varepsilon^* \text{ in } W^{2,m}(\Omega), \text{ and } u_\varepsilon^n \to u_\varepsilon^* \text{ uniformly on } \overline{\Omega} \text{ as } n \to \infty. \tag{4.33}$$

The second assertion follows by the fact that $W^{2,m}(\Omega)$ is compactly embedded in $C(\overline{\Omega})$. By passing to the limit as $n \to \infty$ in (4.28) it follows that u_ε^* is the solution to (4.9) with $v = v_\varepsilon^*$. Because the convergence is uniform, u_ε^* preserves all boundedness properties of u_ε^n, hence $u_\varepsilon^* \in \widetilde{U}$.

Further, by Proposition 4.1 we get that (4.29) has a unique solution

$$\theta_\varepsilon^n, \ y_\varepsilon^n, \ \beta_\varepsilon^*(y_\varepsilon^n) \in W^{1,\infty}([0, T]; L^2(\Omega)) \cap W^{1,2}([0, T]; V) \cap L^\infty(0, T; H^2(\Omega)),$$

satisfying (4.23)–(4.24). Again, $u_\varepsilon^n \in \widetilde{U}$, so $\frac{\theta_0}{u_\varepsilon^n + \varepsilon} \leq y_s$ which implies, because β_ε^* is monotonically increasing, that

$$\beta_\varepsilon^*\left(\frac{\theta_0}{u_\varepsilon^n + \varepsilon}\right) \leq \beta_\varepsilon^*(y_s) = \beta_s^*. \tag{4.34}$$

From here we get

$$\int_\Omega j_\varepsilon \left(\frac{\theta_0}{u_\varepsilon^n + \varepsilon} \right) dx \leq \beta_s^* y_s^* \text{meas}(\Omega). \tag{4.35}$$

By some computations we also deduce that

$$\left\| \beta_\varepsilon^* \left(\frac{\theta_0}{u_\varepsilon^n + \varepsilon} \right) \right\|_V^2 \leq \frac{C}{\varepsilon^4} \|\theta_0\|_{H^2(\Omega)}^2 \beta_M^2(\varepsilon)(\|v_\varepsilon^n\|^2 + 1) \tag{4.36}$$

$$\leq \frac{C}{\varepsilon^4} \|\theta_0\|_{H^2(\Omega)}^2 \beta_M^2(\varepsilon)(d_\varepsilon + 2),$$

and so the right-hand side in (4.23), written for $\beta_\varepsilon^*(y^n)$, becomes independent of n. Consequently, by selecting a subsequence denoted still by the subscript n, we can write the following convergencies, as $n \to \infty$:

$$\beta_\varepsilon^*(y_\varepsilon^n) \overset{w*}{\to} \eta_\varepsilon^* \text{ in } L^\infty(0, T; H^2(\Omega)) \cap W^{1,\infty}([0, T]; L^2(\Omega)), \tag{4.37}$$

$$\beta_\varepsilon^*(y_\varepsilon^n) \rightharpoonup \eta_\varepsilon^* \text{ in } W^{1,2}([0, T]; V),$$

$$y_\varepsilon^n \overset{w*}{\to} y_\varepsilon^* \text{ in } L^\infty(0, T; H^2(\Omega)) \cap W^{1,\infty}([0, T]; L^2(\Omega)), \tag{4.38}$$

$$y_\varepsilon^n \rightharpoonup y_\varepsilon^* \text{ in } W^{1,2}([0, T]; V),$$

$$\theta_\varepsilon^n \overset{w*}{\to} \theta_\varepsilon^* \text{ in } L^\infty(0, T; H^2(\Omega)) \cap W^{1,\infty}([0, T]; L^2(\Omega)), \tag{4.39}$$

$$\theta_\varepsilon^n \rightharpoonup \theta_\varepsilon^* \text{ in } W^{1,2}([0, T]; V).$$

We deduce by the Aubin–Lions theorem that

$$\theta_\varepsilon^n \to \theta_\varepsilon^* \text{ in } L^2(0, T; L^2(\Omega)) \text{ as } n \to \infty, \tag{4.40}$$

and by (4.33), (4.38) and (4.39) we get that

$$\theta_\varepsilon^* = (u_\varepsilon^* + \varepsilon)y_\varepsilon^* \text{ a.e. on } Q. \tag{4.41}$$

The sequence $\{\theta_\varepsilon^n\}_{n \geq 1}$ is bounded in $C([0, T]; L^2(\Omega))$, is equi-continuous and $\|\theta_\varepsilon^n(t)\|_V \leq$ constant. Then, by the Ascoli–Arzelà theorem we obtain that

$$\theta_\varepsilon^n(t) \to \theta_\varepsilon^*(t) \text{ in } L^2(\Omega) \text{ uniformly in } t \in [0, T], \tag{4.42}$$

which implies that $\lim_{n \to \infty} \theta_\varepsilon^n(0) = \theta_\varepsilon^*(0)$. By (4.31) we get

$$(u_\varepsilon^* + \varepsilon)y_\varepsilon^*(0) = \theta_\varepsilon^*(0) = \theta_0. \tag{4.43}$$

We notice that the function $r \to \beta_\varepsilon^*(r)$ is Lipschitz,

$$\|\beta_\varepsilon^*(y_\varepsilon^n) - \beta_\varepsilon^*(y_\varepsilon^*)\| = \int_{y_\varepsilon^*}^{y_\varepsilon^n} \beta_\varepsilon(r)dr \le \frac{\beta_s^* - \beta^*(y_s - \varepsilon)}{\varepsilon} \|y_\varepsilon^n - y_\varepsilon^*\|,$$

so that

$$\beta_\varepsilon^*(y_\varepsilon^n) \to \beta_\varepsilon^*(y_\varepsilon^*) \text{ in } L^2(0,T;L^2(\Omega)) \text{ as } n \to \infty \qquad (4.44)$$

whence $\eta_\varepsilon^* = \beta_\varepsilon^*(y_\varepsilon^*)$ a.e. on Q. Also $r \to K(r)$ is Lipschitz and we have

$$K(y_\varepsilon^n) \to K(y_\varepsilon^*) \text{ in } L^2(0,T;L^2(\Omega)) \text{ as } n \to \infty.$$

Now we can pass to the limit in (4.27) as n goes to ∞, relying on the weakly lower semicontinuity property of the norms. Then

$$d_\varepsilon \le J_\varepsilon(u_\varepsilon^*) \le \liminf_{n\to\infty} J_\varepsilon(u_\varepsilon^n) \le d_\varepsilon$$

so that, $d_\varepsilon = J_\varepsilon(u_\varepsilon^*)$, and u_ε^* is found to be an optimal control. By passing to the limit in (4.30) we still get that

$$\int_Q \frac{d((u_\varepsilon^* + \varepsilon)y_\varepsilon^*)}{dt} \psi dx dt + \int_Q (\nabla\beta_\varepsilon^*(y_\varepsilon^*) - a(x)K(y_\varepsilon^*)) \cdot \nabla\psi dx dt = \int_Q f\psi dx dt,$$

for any $\psi \in L^2(0,T;V)$, which together with (4.43) proves that y_ε^* is the solution to (4.18) with $u = u_\varepsilon^*$. $\qquad \square$

4.3 The Approximating Optimality Condition

In order to compute the optimality condition for an approximating controller u_ε^* we have to introduce and to study the system in variations and the dual system.

4.3.1 The First Order Variations System

Let us denote by u_ε^* a controller and by y_ε^* its corresponding state. Let $w_\varepsilon \in \widetilde{U}$ and $\lambda \in [0,1]$ and define the variation $u_\varepsilon^\lambda = (1-\lambda)u_\varepsilon^* + \lambda w_\varepsilon$ that can be still written

$$u_\varepsilon^\lambda = u_\varepsilon^* + \lambda\widetilde{u_\varepsilon}, \qquad (4.45)$$

where

$$\widetilde{u_\varepsilon} = w_\varepsilon - u_\varepsilon^*. \qquad (4.46)$$

We define by

$$Y_\varepsilon(t,x) = \lim_{\lambda\searrow 0} \frac{y_\varepsilon^\lambda(t,x) - y_\varepsilon^*(t,x)}{\lambda}$$

the Gâteaux derivative of the state function y_ε with respect to the controller u_ε^*. Here, y_ε^λ is the solution to (4.18) corresponding to u_ε^λ. Without indicating all computation we note that formally Y_ε satisfies

$$\frac{\partial}{\partial t}((u_\varepsilon^* + \varepsilon)Y_\varepsilon) - \Delta(\beta_\varepsilon(y_\varepsilon^*)Y_\varepsilon) + \nabla \cdot (a(x)K'(y_\varepsilon^*)Y_\varepsilon) = -\frac{d(\widetilde{u_\varepsilon}y_\varepsilon^*)}{dt} \text{ in } Q,$$
(4.47)

$$Y_\varepsilon(t, x) = 0 \text{ on } \Sigma, \qquad (4.48)$$

$$(u_\varepsilon^* + \varepsilon)Y_\varepsilon(0, x) = -\widetilde{u_\varepsilon}y_\varepsilon^*(0, x) \text{ in } \Omega.$$
(4.49)

For $u_\varepsilon^* \in \widetilde{U}$ we set

$$v_\varepsilon^* = u_\varepsilon^* - \Delta u_\varepsilon^*, \quad v_\varepsilon^\lambda = u_\varepsilon^\lambda - \Delta u_\varepsilon^\lambda, \qquad (4.50)$$

define

$$\widetilde{v}_\varepsilon(t, x) = \lim_{\lambda \searrow 0} \frac{v_\varepsilon^\lambda(t, x) - v_\varepsilon^*(t, x)}{\lambda}$$

and deduce that

$$\widetilde{v}_\varepsilon = \widetilde{u}_\varepsilon - \Delta\widetilde{u}_\varepsilon \text{ in } \Omega. \qquad (4.51)$$

We also note that $\nabla\widetilde{u}_\varepsilon \cdot \nu = 0$ and $\nabla u_\varepsilon^* \cdot \nu = 0$ on Γ because $u_\varepsilon^*, \widetilde{u}_\varepsilon \in \widetilde{U}$.

Proposition 4.3. *Assume (4.5) and (4.6). Then, problem (4.47)–(4.49) has a unique solution*

$$Y_\varepsilon \in C([0, T]; L^2(\Omega)) \cap W^{1,2}([0, T]; V') \cap L^2(0, T; V). \qquad (4.52)$$

Proof. We denote $z_\varepsilon = (u_\varepsilon^* + \varepsilon)Y_\varepsilon$, and write the abstract linear Cauchy problem (equivalent to (4.47)–(4.49))

$$\frac{dz_\varepsilon}{dt}(t) + B_\varepsilon^Y(t)z_\varepsilon(t) = F_\varepsilon^Y(t) \text{ a.e. } t \in (0, T), \qquad (4.53)$$

$$z_\varepsilon(0) = z_{0\varepsilon}, \qquad (4.54)$$

where

$$F_\varepsilon^Y = -\frac{d(\widetilde{u_\varepsilon}y_\varepsilon^*)}{dt} \in L^2(0, T; L^2(\Omega)), \ z_{0\varepsilon} = -\widetilde{u_\varepsilon}y_\varepsilon^*(0) \in V,$$

and $B_\varepsilon^Y(t) : V \to V'$ is the time-dependent linear operator defined by

$$\langle B_\varepsilon^Y(t)z, \psi\rangle_{V',V} = \int_\Omega (\nabla(g(t)z) - b(x)K'(y_\varepsilon^*)z) \cdot \nabla\psi dx, \text{ for any } \psi \in V,$$

$$(4.55)$$

with

$$g(t,x) = \frac{\beta_\varepsilon(y_\varepsilon^*(t,x))}{u_\varepsilon^*(x) + \varepsilon}, \quad b(x) = \frac{1}{u_\varepsilon^*(x) + \varepsilon}a(x). \tag{4.56}$$

First of all, we recall that $0 \le \rho \le \beta_\varepsilon(y_\varepsilon^*(t)) \le \beta_M(\varepsilon)$ given by (4.25). By (4.12), $\beta_\varepsilon' \in C^1(\mathbb{R})$, $\beta_\varepsilon'(r) = 0$ for $r \in \mathbb{R}\backslash[0, y_s]$, so $\beta_\varepsilon'(r)$ is bounded on $[0, y_s]$ by a constant dependent on ε. Also, $\varepsilon \le u_\varepsilon^*(x) + \varepsilon \le u_M + 1$, so we have that

$$|\nabla g| \le \frac{|(u_\varepsilon^* + 1)\beta_\varepsilon'(y_\varepsilon^*)\nabla y_\varepsilon^*| + |\beta_\varepsilon(y_\varepsilon^*)\nabla u_\varepsilon^*|}{\varepsilon^2}, \quad |b| \le \frac{\|a\|_{L^\infty(\Omega)}}{2\varepsilon}.$$

By (4.24) we get that $g \in L^\infty(0, T; H^2(\Omega)) \cap L^\infty(Q)$,

$$g(t,x) \ge \frac{\rho}{u_M + 1}, \quad \|\nabla g\| \le \frac{C(\varepsilon, \|v_\varepsilon^*\|_{L^m(\Omega)})}{\varepsilon^2} = C(\varepsilon), \tag{4.57}$$

where ∇ denotes the operator $\nabla = \left(\frac{\partial}{\partial x_1}, ..., \frac{\partial}{\partial x_N}\right)$ and $C(\varepsilon)$ accounts for several constants depending on ε.

The function $t \to B_\varepsilon^Y(t)z$ is measurable from $[0, T]$ to V', and we shall deduce some other properties of $B_\varepsilon^Y(t)$. To this end we use the results below established for $N \le 3$.

If $\phi \in H_0^1(\Omega) \subset L^6(\Omega)$ we have $\phi^2 \in L^2(\Omega)$. Indeed, by the Sobolev embedding theorem we get

$$\|\phi^2\|^2 = \int_\Omega \phi^4 dx \le \left(\int_\Omega \phi^2 dx\right)^{1/2}\left(\int_\Omega \phi^6 dx\right)^{1/2} \tag{4.58}$$

$$= \|\phi\| \|\phi\|_{L^6(\Omega)}^3 \le C \|\phi\| \|\phi\|_V^3.$$

Hence, if $\phi, \psi \in V$ it follows that $\phi\psi \in L^2(\Omega)$ and

$$\|\phi\psi\| = \left(\int_\Omega \phi^2\psi^2 dx\right)^{1/2} \le C \|\phi\|^{1/2} \|\phi\|_V^{3/2} \|\psi\|^{1/2} \|\psi\|_V^{3/2}. \tag{4.59}$$

Next we compute

$$\langle B_\varepsilon^Y(t)z, z\rangle_{V',V} = \int_\Omega \left(g(t,x)|\nabla z|^2 + z(\nabla g(t,x) - b(x)K'(y_\varepsilon^*)) \cdot \nabla z\right) dx$$

$$\ge \frac{\rho}{2(u_M + 1)} \|z\|_V^2 - \frac{(u_M + 1)}{2\rho} \|z(\nabla g - bK'(y_\varepsilon^*))\|^2.$$

Applying (4.59) and (4.57) we get

$$\|z(\nabla g - bK'(y_\varepsilon^*))\|^2$$
$$\leq C \|z\|^{1/2} \|z\|_V^{3/2} \|\nabla g - bK'(y_\varepsilon^*)\|^{1/2} \|\nabla g - bK'(y_\varepsilon^*)\|_V^{3/2}$$
$$\leq C(\varepsilon) \|z\|^{1/2} \|z\|_V^{3/2}$$

and obtain, by a consequence of Hölder's theorem, that

$$\langle B_\varepsilon^Y(t)z, z\rangle_{V',V} \geq \frac{\rho}{4(u_M + 1)} \|z\|_V^2 - C(\varepsilon) \|z\|^2 .$$

Again by (4.57) we have

$$\left|\langle B_\varepsilon^Y(t)z, \psi\rangle_{V',V}\right| = \int_\Omega (g(t,x)\nabla z + z\nabla g(t,x) - b(x)K'(y_\varepsilon^*)z) \cdot \nabla\psi) \, dx$$
$$\leq C(\varepsilon) \|z\|_V \|\psi\|_V ,$$

hence

$$\left\|B_\varepsilon^Y(t)z\right\|_{V'} \leq C(\varepsilon) \|z\|_V .$$

In conclusion $B_\varepsilon^Y(t)$ satisfies the properties required by the theorem of Lions (see [77]) to ensure the existence and uniqueness of the solution to (4.47)–(4.49), i.e., $z_\varepsilon \in W^{1,2}([0,T]; V') \cap L^2(0,T; V)$. We turn back to the function Y_ε and get (4.52). □

4.3.2 The Dual System

We denote by $\chi_{(0,T_{obs})}$, $\chi_{\Omega_{obs}}$, χ_{Ω_0} the characteristic functions of $(0, T_{obs})$, Ω_{obs} and Ω_0 respectively, by p_ε the dual variable and introduce the dual system

$$\frac{\partial}{\partial t}((u_\varepsilon^* + \varepsilon)p_\varepsilon) + \beta_\varepsilon(y_\varepsilon^*)\Delta p_\varepsilon + a(x)K'(y_\varepsilon^*) \cdot \nabla p_\varepsilon = E_\varepsilon \text{ in } Q, \qquad (4.60)$$

$$p_\varepsilon(t,x) = 0 \text{ on } \Sigma, \qquad (4.61)$$

$$p_\varepsilon(T,x) = 0 \text{ in } \Omega, \qquad (4.62)$$

where

$$E_\varepsilon(t,x) = -(u_\varepsilon^*(x) + \varepsilon)S_\varepsilon^*(t,x)\chi_{(0,T_{obs})}(t)\chi_{\Omega_{obs}}(x), \qquad (4.63)$$

$$S_\varepsilon^*(t,x) = (u_\varepsilon^* + \varepsilon)y_\varepsilon^*(t,x) - \theta_{obs}(t,x),$$

$$S_\varepsilon^* \in L^2(0,T; L^2(\Omega)), \ E_\varepsilon \in L^2(0,T; L^2(\Omega)).$$

Making the transformation $t = T - \tau$ we have to study the abstract linear Cauchy problem

$$\frac{d((u_\varepsilon^* + \varepsilon)\widetilde{p}_\varepsilon)}{d\tau}(\tau) + G_\varepsilon(\tau)\widetilde{p}_\varepsilon(\tau) = \widetilde{E}_\varepsilon(\tau) \text{ a.e. } \tau \in (0,T), \qquad (4.64)$$

$$\widetilde{p}_\varepsilon(0,x) = 0, \qquad (4.65)$$

where $\widetilde{p}_\varepsilon(\tau,x) = p_\varepsilon(T-\tau,x)$, $\widetilde{E}_\varepsilon(\tau,x) = E_\varepsilon(T-\tau,x)$, $G_\varepsilon(\tau) : V \to V'$ is the time-dependent linear operator

$$\langle G_\varepsilon(\tau)z, \psi \rangle_{V',V} = \int_\Omega \{g(t,x)\nabla\psi + \psi(\nabla g(t,x) - b(x)K'(y_\varepsilon^*))\} \cdot \nabla z dx \qquad (4.66)$$

for any $\psi \in V$, with g and b previously defined in (4.56).

Proposition 4.4. *Assume (4.5) and (4.6). Then, problem (4.64)–(4.65) has a unique solution*

$$p_\varepsilon \in C([0,T]; L^2(\Omega)) \cap W^{1,2}([0,T]; V') \cap L^2(0,T; V). \qquad (4.67)$$

The proof is led in a similar way as for the system in variations.

4.3.3   The Necessary Optimality Condition

Proposition 4.5. *Assume (4.5)–(4.6) and let $(u_\varepsilon^*, y_\varepsilon^*)$ be an optimal pair in (P_ε). Then, the necessary optimality condition has the variational form*

$$\int_\Omega (u_\varepsilon^* - w_\varepsilon)\alpha_\varepsilon dx - k_1 \int_\Omega (v_\varepsilon^*)^{m-1}\Delta(u_\varepsilon^* - w_\varepsilon)dx \leq 0, \qquad (4.68)$$

where $w_\varepsilon \in \widetilde{U}$, $v_\varepsilon^* = u_\varepsilon^* - \Delta u_\varepsilon^*$ and

$$\alpha_\varepsilon(x) = -p_\varepsilon(0,x)y_\varepsilon^*(0,x) \qquad (4.69)$$

$$-\int_0^T \left(p_\varepsilon \frac{dy_\varepsilon^*}{dt} - \chi_{(0,T_{obs})}\chi_{\Omega_{obs}}y_\varepsilon^*S_\varepsilon^* \right) dt + \frac{1}{\varepsilon}u_\varepsilon^*\chi_{\Omega_0} + k_1(v_\varepsilon^*)^{m-1}.$$

Proof. We multiply (4.47) by p_ε and integrate over Q,

$$-\int_0^T \left\langle \frac{d}{dt}((u_\varepsilon^* + \varepsilon)p_\varepsilon)(t) + \beta_\varepsilon(y_\varepsilon^*(t))\Delta p_\varepsilon(t), Y_\varepsilon(t) \right\rangle_{V',V} dt \quad (4.70)$$

$$+ \int_\Omega (u_\varepsilon^* + \varepsilon)(Y_\varepsilon p_\varepsilon)(T, x)dx - \int_\Omega (u_\varepsilon^* + \varepsilon)(Y_\varepsilon p_\varepsilon)(0, x)dx$$

$$+ \int_\Sigma Y_\varepsilon \beta_\varepsilon(y_\varepsilon^*) \frac{\partial p_\varepsilon}{\partial \nu} d\sigma dt - \int_\Sigma p_\varepsilon \frac{\partial}{\partial \nu}(\beta_\varepsilon(y_\varepsilon^*) Y_\varepsilon)d\sigma dt$$

$$= - \int_Q \widetilde{u_\varepsilon} \frac{dy_\varepsilon^*}{dt} p_\varepsilon dx dt,$$

where $\widetilde{u_\varepsilon}$ is defined in (4.46). Taking into account the boundary conditions for Y_ε and $\widetilde{u_\varepsilon}$ and the initial condition for Y_ε we obtain

$$\int_Q Y_\varepsilon S_\varepsilon^*(u_\varepsilon^* + \varepsilon)\chi_{(0,T_{obs})}\chi_{\Omega_{obs}} dx dt \quad (4.71)$$

$$= - \int_\Omega p_\varepsilon(0, x)y_\varepsilon^*(0, x)\widetilde{u_\varepsilon} dx - \int_Q p_\varepsilon \frac{dy_\varepsilon^*}{dt}\widetilde{u_\varepsilon} dx dt.$$

On the other hand, u_ε^* is optimal in (P_ε), so that is verifies

$$\frac{1}{2}\int_{Q_{obs}} ((u_\varepsilon^* + \varepsilon)y_\varepsilon^* - \theta_{obs})^2 dx dt + \frac{k_1}{m}\int_\Omega (v_\varepsilon^*)^m dx + \frac{1}{2\varepsilon}\int_{\Omega_0} (u_\varepsilon^*)^2 dx$$

$$\leq \frac{1}{2}\int_{Q_{obs}} ((u_\varepsilon^\lambda + \varepsilon)y_\varepsilon^\lambda - \theta_{obs})^2 dx dt + \frac{k_1}{m}\int_\Omega (v_\varepsilon^\lambda)^m dx + \frac{1}{2\varepsilon}\int_{\Omega_0} (u_\varepsilon^\lambda)^2 dx,$$

where we recall that u_ε^λ and v_ε^λ were defined in (4.45) and (4.50). Therefore, by a few computation we obtain

$$\int_{Q_{obs}} ((u_\varepsilon^* + \varepsilon)Y_\varepsilon + \widetilde{u_\varepsilon}y_\varepsilon^*)S_\varepsilon^* dx dt + k_1\int_\Omega (v_\varepsilon^*)^{m-1}\widetilde{v_\varepsilon} dx + \frac{1}{\varepsilon}\int_{\Omega_0} u_\varepsilon^*\widetilde{u_\varepsilon} dx \geq 0.$$

$$(4.72)$$

We replace (4.71) in the left-hand side of (4.72), use (4.51) and get

$$\int_\Omega \widetilde{u_\varepsilon}\alpha_\varepsilon dx - k_1\int_\Omega (v_\varepsilon^*)^{m-1}\Delta\widetilde{u_\varepsilon} dx \geq 0,$$

which implies (4.68) as claimed.   □

4.4 Convergence of the Approximating Control Problem

Now we are going to prove that a sequence $(u_\varepsilon^*, y_\varepsilon^*)_\varepsilon$ of the optimal pairs in (P_ε) converges as ε goes to 0 to a pair (u^*, y^*) that realizes the minimum in (P). Moreover, the function y^* turns out to be the solution to (4.1)–(4.3). Before this we give a preliminary result.

Let us write again the abstract Cauchy problem (1.27) equivalent to (4.1)–(4.3)

$$\frac{d(uy)}{dt}(t) + Ay(t) \ni f(t) \text{ a.e. } t \in (0, T), \tag{4.73}$$

$$(uy(t))|_{t=0} = \theta_0,$$

where $u = 0$ on $\overline{\Omega_0}$, and A was defined in Sect. 1.1, i.e.,

$$A : D(A) \subset V' \to V',$$

$$D(A) = \{y \in L^2(\Omega); \text{there exists } \zeta \in V, \ \zeta \in \beta^*(y) \text{ a.e. on } \Omega\},$$

$$\langle Ay, \psi \rangle_{V',V} = \int_\Omega (\nabla\zeta - a(x)K(y)) \cdot \nabla\psi dx, \text{ for any } \psi \in V,$$

and $\zeta \in \beta^*(y)$ a.e. on Ω.

Lemma 4.6. *Let (4.5) and (4.6) hold and assume*

$$u_\varepsilon \in W^{2,m}(\Omega), \ 0 \leq \frac{\theta_0}{y_s} \leq u_\varepsilon \leq u_M \text{ on } \Omega, \tag{4.74}$$

such that

$$u_\varepsilon \rightharpoonup u \text{ in } W^{2,m}(\Omega), \text{ as } \varepsilon \to 0, \ u = 0 \text{ on } \overline{\Omega_0}. \tag{4.75}$$

Let y_ε be the solution to the approximating state system (4.18). Then, there exists a subsequence of $(y_\varepsilon)_\varepsilon$ denoted still by the subscript ε, such that

$$y_\varepsilon \rightharpoonup y \text{ in } L^2(0, T; V), \text{ as } \varepsilon \to 0, \tag{4.76}$$

$$(u_\varepsilon + \varepsilon)y_\varepsilon = \theta_\varepsilon \rightharpoonup \theta = uy \text{ in } W^{1,2}([0, T]; V') \cap L^2(0, T; V), \tag{4.77}$$

$$\theta_\varepsilon \to \theta \text{ strongly in } L^2(0, T; L^2(\Omega)), \text{ as } \varepsilon \to 0,$$

and y is a solution to (4.73) corresponding to u.

Proof. Let u_ε satisfy (4.74)–(4.75). Then, by Proposition 4.1 we know that (4.18) has a unique solution y_ε with the properties (4.20)–(4.24).

Since $u_\varepsilon + \varepsilon \geq u_\varepsilon$ on Ω we have $\frac{\theta_0}{u_\varepsilon + \varepsilon} \leq y_s$, which implies

$$\beta_\varepsilon^* \left(\frac{\theta_0}{u_\varepsilon + \varepsilon} \right) \leq \beta_\varepsilon^*(y_s) = \beta_s^*, \quad \int_\Omega j_\varepsilon \left(\frac{\theta_0}{u_\varepsilon + \varepsilon} \right) dx \leq \beta_s^* y_s \text{meas}(\Omega),$$

and so the right-hand side in (4.21) becomes independent of ε. Therefore by selecting a subsequence, if necessary, we get

$$\beta_\varepsilon^* (y_\varepsilon) \rightharpoonup \zeta \text{ in } L^2(0,T;V) \text{ as } \varepsilon \to 0,$$

$$y_\varepsilon \rightharpoonup y \text{ in } L^2(0,T;V) \text{ as } \varepsilon \to 0.$$

By (4.75) we have

$$u_\varepsilon \to u \text{ uniformly in } \overline{\Omega} \text{ as } \varepsilon \to 0,$$

then

$$(u_\varepsilon + \varepsilon)y_\varepsilon = \theta_\varepsilon \rightharpoonup \theta \text{ in } L^2(0,T;L^2(\Omega)) \text{ as } \varepsilon \to 0$$

and so we get that $\theta = uy$ a.e. on Q. Again by (4.21) we have

$$\frac{d\theta_\varepsilon}{dt} \rightharpoonup \frac{d\theta}{dt} \text{ in } L^2(0,T;V') \text{ as } \varepsilon \to 0.$$

We deduce via the Aubin–Lions theorem and Ascoli–Arzelà theorem, respectively, that

$$\theta_\varepsilon \to \theta \text{ in } L^2(0,T;L^2(\Omega)) \text{ as } \varepsilon \to 0,$$

$$\theta_\varepsilon(t) \to \theta(t) \text{ in } V' \text{ uniformly for } t \in [0,T], \text{ as } \varepsilon \to 0,$$

which implies

$$\lim_{n \to \infty} \theta_\varepsilon(0) = \theta(0) = \theta_0.$$

We still remark, for a further use, that

$$(u_\varepsilon + \varepsilon)y_\varepsilon - \theta_{obs} \to uy - \theta_{obs} \text{ in } L^2(0,T;L^2(\Omega)) \text{ as } \varepsilon \to 0. \tag{4.78}$$

Now we have all ingredients to prove that y is a solution to (4.73) (equivalently (4.1)–(4.3)) by applying Theorem 1.6 in Sect. 1.1.   □

Theorem 4.7. *Let (4.5)–(4.6) hold, and let $(u_\varepsilon^*, y_\varepsilon^*)$ be optimal in (P_ε). Then, there exists a subsequence denoted still by the subscript ε, such that*

$$u_\varepsilon^* \rightharpoonup u^* \text{ in } W^{2,m}(\Omega), \text{ as } \varepsilon \to 0, \tag{4.79}$$

$$y_\varepsilon^* \rightharpoonup y^* \text{ in } L^2(0,T;V), \text{ as } \varepsilon \to 0, \tag{4.80}$$

$$(u_\varepsilon^* + \varepsilon)y_\varepsilon^* = \theta_\varepsilon^* \to \theta^* = u^* y^* \text{ in } L^2(0,T;L^2(\Omega)) \text{ as } \varepsilon \to 0, \quad (4.81)$$

$$\theta_\varepsilon^* \to \theta^* \text{ in } W^{1,2}([0,T];V') \cap L^2(0,T;V).$$

Moreover, $u^* \in U$, y^* is a solution to (4.1)–(4.3) and (u^*, y^*) realizes the minimum in (P), i.e.,

$$\frac{1}{2} \int_{Q_{obs}} (u^*(x)y^*(t,x) - \theta_{obs}(t,x))^2 dx dt + \frac{k_1}{m} \int_\Omega (u^* - \Delta u^*)^m (x) dx$$

$$\leq \frac{1}{2} \int_{Q_{obs}} (u(x)y(t,x) - \theta_{obs}(t,x))^2 dx dt + \frac{k_1}{m} \int_\Omega (u - \Delta u)^m (x) dx, \quad (4.82)$$

for any $u \in U$, and y a solution to (4.1)–(4.3).

Proof. Let $(u_\varepsilon^*, y_\varepsilon^*)$ be optimal in (P_ε) and denote $v_\varepsilon^* = u_\varepsilon^* - \Delta u_\varepsilon^*$ a.e. on Ω. By the optimality of $(u_\varepsilon^*, y_\varepsilon^*)$ we can write

$$\frac{1}{2} \int_{Q_{obs}} ((u_\varepsilon^* + \varepsilon)y_\varepsilon^* - \theta_{obs})^2 dx dt + \frac{k_1}{m} \int_\Omega (v_\varepsilon^*)^m dx + \frac{1}{2\varepsilon} \int_{\Omega_0} (u_\varepsilon^*)^2 dx$$

$$\leq \frac{1}{2} \int_{Q_{obs}} ((u + \varepsilon)y_\varepsilon - \theta_{obs})^2 dx dt + \frac{k_1}{m} \int_\Omega v^m dx + \frac{1}{2\varepsilon} \int_{\Omega_0} u^2 dx, \quad (4.83)$$

for any $u \in \tilde{U}$, where y_ε is the solution to the approximating state system (4.18) corresponding to u. We recall that U and $\tilde{U}$ are the admissible sets for (P) and (P_ε), described by (4.7) and (4.14). In particular, let us take $u \in U \subset \tilde{U}$. We remark that the integral on Ω_0 vanishes on the right-hand side because $u \in U$ vanishes on $\overline{\Omega_0}$ and $v = u - \Delta u$ is in $L^2(\Omega)$.

We apply Lemma 4.6 with $u_\varepsilon = u$. Thus, y_ε tends to y which is a solution to (4.73) (equivalently (4.1)–(4.3)), according to the same convergencies as in (4.76)–(4.78). Since the last term on the right-hand side in (4.83) is zero, we can write on the basis of the strong convergence (4.78),

$$\frac{1}{2} \int_{Q_{obs}} ((u + \varepsilon)y_\varepsilon - \theta_{obs})^2 dx dt + \frac{k_1}{m} \int_\Omega v^m dx \quad (4.84)$$

$$\leq \limsup_{\varepsilon \to 0} \left\{ \frac{1}{2} \int_{Q_{obs}} ((u + \varepsilon))y_\varepsilon - \theta_{obs})^2 dx dt + \frac{k_1}{m} \int_\Omega v^m dx \right\}$$

$$\leq \frac{1}{2} \int_{Q_{obs}} (uy - \theta_{obs})^2 dx dt + \frac{k_1}{m} \int_\Omega v^m dx = \text{constant}.$$

Since the left-hand side in (4.83) is bounded by a constant selecting a subsequence, if necessary, we get

$$v_\varepsilon^* \rightharpoonup v^* \text{ in } L^m(\Omega), \text{ as } \varepsilon \to 0,$$

$$\frac{1}{\sqrt{\varepsilon}} u_\varepsilon^* \rightharpoonup u_0 \text{ in } L^2(\Omega_0), \text{ as } \varepsilon \to 0. \qquad (4.85)$$

We recall that $\dot{u}_\varepsilon^*$ satisfies (4.9) and so

$$u_\varepsilon^* \rightharpoonup u^* \text{ in } W^{2,m}(\Omega), \text{ and uniformly on } \overline{\Omega}, \text{ as } \varepsilon \to 0. \qquad (4.86)$$

Moreover u^* is the solution to (4.9) corresponding to v^*, i.e., $v^* = u^* - \Delta u^*$ a.e. on Ω. By (4.85) we derive that

$$u^* = w\text{-}\lim_{\varepsilon \to 0} u_\varepsilon^* = 0 \text{ in } \Omega_0. \qquad (4.87)$$

Consequently, $u^* \in U$. Again by Lemma 4.6 with u_ε replaced by u_ε^* we deduce that y^* is a solution to (4.1)–(4.3) corresponding to u^*. On the basis of these results and taking into account that the integral on Ω_0 is nonnegative, we can write successively from (4.83) and (4.84) that

$$\frac{1}{2} \int_{Q_{obs}} (u^* y^* - \theta_{obs})^2 dx dt + \frac{k_1}{m} \int_\Omega (v^*)^m dx$$

$$\leq \liminf_{\varepsilon \to 0} \left\{ \frac{1}{2} \int_{Q_{obs}} ((u_\varepsilon^* + \varepsilon) y_\varepsilon^* - \theta_{obs})^2 dx dt + \frac{k_1}{m} \int_\Omega (v_\varepsilon^*)^m dx \right.$$

$$\left. + \frac{1}{2\varepsilon} \int_{\Omega_0} (u_\varepsilon^*)^2 dx \right\}$$

$$\leq \liminf_{\varepsilon \to 0} \left\{ \frac{1}{2} \int_{Q_{obs}} ((u + \varepsilon) y_\varepsilon - \theta_{obs})^2 dx dt + \frac{k_1}{m} \int_\Omega v^m dx \right\}$$

$$\leq \limsup_{\varepsilon \to 0} \left\{ \frac{1}{2} \int_{Q_{obs}} ((u + \varepsilon) y_\varepsilon - \theta_{obs})^2 dx dt + \frac{k_1}{m} \int_\Omega v^m dx \right\}$$

$$\leq \frac{1}{2} \int_{Q_{obs}} (uy - \theta_{obs})^2 dx dt + \frac{k_1}{m} \int_\Omega v^m dx,$$

for any $u \in U$ and this proves that (u^*, y^*) realizes the minimum in (P). □

4.5 An Alternative Approach

We present now another approach of the control problem (P), based on the control change. The advantage is that the optimality conditions will be found in a simpler form, but an additional restriction will be required for the initial data θ_0. More exactly, we assume (4.5) and

$$\theta_0 \in H^2(\Omega) \cap H^1_0(\Omega) \subset C(\overline{\Omega}), \ \theta_0 = 0 \ \text{in} \ \Omega_0, \tag{4.88}$$

$$\theta_0 \geq 0 \ \text{in} \ \Omega_u, \ \frac{\theta_0}{u} \in L^2(\Omega_u), \ \frac{\theta_0}{u} \leq y_s, \ \text{for} \ x \in \Omega_u,$$

$$\frac{\partial \theta_0}{\partial \nu} \leq 0 \ \text{on} \ \Gamma.$$

Let us denote

$$\phi_{\min} = \frac{1}{y_s} (\theta_0 - \Delta \theta_0), \tag{4.89}$$

and consider problem (4.9). We prove

Lemma 4.8. *Let*

$$v \in L^m(\Omega), \ \phi_{\min}(x) \leq v \leq u_M \ \text{a.e. on} \ \Omega. \tag{4.90}$$

Then, the unique solution $u \in W^{2,m}(\Omega)$ to problem (4.9) satisfies

$$\frac{\theta_0(x)}{y_s} \leq u(x) \leq u_M, \ \text{for any} \ x \in \Omega. \tag{4.91}$$

Proof. To show the lower boundedness of u we denote $z = u - \frac{\theta_0}{y_s}$ and write the problem

$$z - \Delta z = v - \phi_{\min} \ \text{in} \ \Omega, \tag{4.92}$$

$$\nabla z \cdot \nu = -\frac{\partial \theta_0}{\partial \nu} \ \text{on} \ \Gamma.$$

We multiply (4.92) by z^- (the negative part of z) and integrate over Ω, applying the Stampacchia's lemma. We have

$$\left\| z^- \right\|^2_{H^1(\Omega)} = - \int_\Omega (v - \phi_{\min}) z^- dx + \int_{\partial \Omega} \frac{\partial \theta_0}{\partial \nu} z^- d\sigma \leq 0,$$

since $v \geq \phi_{\min}$ and $\frac{\partial \theta_0}{\partial \nu} \leq 0$ on Γ by (4.88). It follows that $u(x) \geq \frac{\theta_0}{y_s} \geq 0$ in Ω. The inequality $u(x) \leq u_M$ is similarly shown. $\square$

Let us fix $m > N$ and consider the new control problem

$$\text{Minimize} \ \left\{ \int_{Q_{obs}} (u(x) y(t, x) - \theta_{obs}(t, x))^2 dx dt \right\} \tag{$\tilde{P}$}$$

subject to (4.1)–(4.3), for all $(v, u) \in W$

$$W = \{(v, u); \ v \in L^\infty(\Omega), \ \phi_{\min} \leq v \leq u_M \ \text{a.e. on} \ \Omega, \tag{4.93}$$

$$u \ \text{satisfies (4.9), and} \ u = 0 \ \text{in} \ \overline{\Omega_0}\}.$$

In this problem we have two controls u and v which are related by (4.9), where $v \in L^\infty(\Omega) \subset L^m(\Omega)$, with $m > N$.

4.5.1 The Approximating Problem $(\widetilde{P}_\varepsilon)$

Let us denote

$$\widetilde{W} = \{v \in L^\infty(\Omega); \ \phi_{\min}(x) \leq v \leq u_M \ \text{ a.e. on } \Omega\} \tag{4.94}$$

and introduce the problem

$$\text{Minimize} \left\{ \int_{Q_{obs}} ((u^v(x) + \varepsilon)y(t,x) - \theta_{obs}(t,x))^2 dxdt + \frac{1}{\varepsilon} \int_{\Omega_0} (u^v)^2(x)dx \right\}$$
$$(\widetilde{P}_\varepsilon)$$

for all $v \in \widetilde{W}$, subject to the approximating problem (4.15)–(4.17), where u^v is the solution to (4.9) with $v \in \widetilde{W}$. We denote

$$\widetilde{J}_\varepsilon(v) = \int_{Q_{obs}} ((u^v(x) + \varepsilon)y(t,x) - \theta_{obs}(t,x))^2 dxdt + \frac{1}{\varepsilon} \int_{\Omega_0} (u^v)^2(x)dx. \tag{4.95}$$

In $(\widetilde{P}_\varepsilon)$ and $\widetilde{J}_\varepsilon$ we have written u^v, in order to stress that u^v is determined by v, but further we shall skip this notation for the writing simplicity.

We notice that in $(\widetilde{P}_\varepsilon)$ the only control which remains is v while u becomes a state, being computed by (4.9). Existence of an optimal pair is proved similarly as in Theorem 4.2.

Proposition 4.9. *Assume (4.5) and (4.88). Then, problem $(\widetilde{P}_\varepsilon)$ has at least one solution $v_\varepsilon^* \in \widetilde{W}$.*

Proof. Let ε be fixed. Since $\widetilde{J}_\varepsilon(v) \geq 0$, it follows that $d_\varepsilon = \inf_{v \in \widetilde{W}} \widetilde{J}_\varepsilon(v)$ exists and it is nonnegative. Let $(v_\varepsilon^n)_{n \geq 1}$ be a minimizing sequence with $v_\varepsilon^n \in \widetilde{W}$. Then

$$d_\varepsilon \leq \int_{Q_{obs}} ((u_\varepsilon^n + \varepsilon)y_\varepsilon^n(t,x) - \theta_{obs}(t,x))^2 dxdt + \frac{1}{\varepsilon} \int_{\Omega_0} (u_\varepsilon^n)^2(x)dx \leq d_\varepsilon + \frac{1}{n}$$

where u_ε^n is the solution to

$$u_\varepsilon^n - \Delta u_\varepsilon^n = v_\varepsilon^n \text{ in } \Omega, \tag{4.96}$$

$$\frac{\partial u_\varepsilon^n}{\partial \nu} = 0 \text{ on } \Gamma$$

and y_ε^n is the solution to (4.18) corresponding to u_ε^n, i.e.,

$$\frac{d((u_\varepsilon^n + \varepsilon)y_\varepsilon^n)}{dt}(t) + A_\varepsilon y_\varepsilon^n(t) = f(t) \text{ a.e. } t \in (0, T),$$

$$u_\varepsilon^n y_\varepsilon^n(0) = \theta_0.$$

But $v_\varepsilon^n \in \widetilde{W}$, so $\|v_\varepsilon^n\|_{L^m(\Omega)} \leq$ constant and it follows that on a subsequence $v_\varepsilon^n \rightharpoonup v_\varepsilon^*$ as $n \to \infty$. The boundedness of u_ε^n follows by (4.96) and everything continues like in Theorem 4.2. $\qquad \square$

Before establishing the form of the new optimality condition we specify that the convergence result remains true.

Theorem 4.10. *Let (4.5), (4.88) hold, and let $(v_\varepsilon^*, y_\varepsilon^*)$ be optimal in $(\widetilde{P_\varepsilon})$. Then, there exists a subsequence denoted still by the subscript ε, such that*

$$v_\varepsilon^* \rightharpoonup v^* \text{ in } L^2(\Omega), \text{ as } \varepsilon \to 0,$$

$$u_\varepsilon^* \rightharpoonup u^* \text{ in } W^{2,m}(\Omega), \text{ as } \varepsilon \to 0,$$

$$y_\varepsilon^* \rightharpoonup y^* \text{ in } L^2(0, T; V), \text{ as } \varepsilon \to 0,$$

$$(u_\varepsilon^* + \varepsilon)y_\varepsilon^* = \theta_\varepsilon^* \to \theta^* = u^* y^* \text{ in } L^2(0, T; L^2(\Omega)) \text{ as } \varepsilon \to 0,$$

$$\theta_\varepsilon^* \to \theta^* \text{ in } W^{1,2}([0, T]; V') \cap L^2(0, T; V).$$

Moreover, $u^* \in U$, y^* is a solution to (4.1)–(4.3) and $((u^*, v^*), y^*)$ realizes the minimum in $(\widetilde{P})$, i.e.,

$$\int_{Q_{obs}} (u^*(x)y^*(t, x) - \theta_{obs}(t, x))^2 dx dt \leq \int_{Q_{obs}} (u(x)y(t, x) - \theta_{obs}(t, x))^2 dx dt,$$

for any $(u, v) \in W$, and y a solution to (4.1)–(4.3).

Proof. Let $(v_\varepsilon^*, y_\varepsilon^*)$ be optimal in $(\widetilde{P_\varepsilon})$ and let u_ε^* be the function given by the elliptic equation. By the optimality of $(v_\varepsilon^*, y_\varepsilon^*)$ we can write

$$\int_{Q_{obs}} (u_\varepsilon^* y_\varepsilon^* - \theta_{obs})^2 dx dt + \frac{1}{\varepsilon} \int_{\Omega_0} (u_\varepsilon^*)^2 dx \leq \int_{Q_{obs}} (u y_\varepsilon - \theta_{obs})^2 dx dt + \frac{1}{\varepsilon} \int_{\Omega_0} u^2 dx,$$

for any $v \in \widetilde{W}$, where y_ε is the solution to the approximating state system (4.18) corresponding to u via v. In particular, let us take v as an element of a pair $(u, v) \in W$. We remark that the integral on Ω_0 vanishes on the right-hand side and next one can continue as in Theorem 4.7. $\qquad \square$

4.5.1.1   The First Order Variations and Dual Systems

Assume that v_ε^* is a controller and u_ε^* and y_ε^* are the corresponding states. Let $w_\varepsilon \in \widetilde{W}$ and $\lambda \in [0, 1]$ and denote the variation along the direction λ by $v_\varepsilon^\lambda = v_\varepsilon^* + \lambda \widetilde{v}_\varepsilon$, where $\widetilde{v}_\varepsilon = w_\varepsilon - v_\varepsilon^*$. The optimal state u_ε^* is given by (4.9) for $v = v_\varepsilon^*$ and $\widetilde{u}_\varepsilon$ is computed from the system

$$\widetilde{u}_\varepsilon - \Delta \widetilde{u}_\varepsilon = \widetilde{v}_\varepsilon \text{ in } \Omega, \tag{4.97}$$

$$\nabla \widetilde{u}_\varepsilon \cdot \nu = 0 \text{ on } \Gamma.$$

The first order variation system (4.47)–(4.49) and the dual system (4.60)–(4.62) are the same as before. Because two states are involved, we have to write a dual system for the state u_ε^*, too. This reads

$$q_\varepsilon - \Delta q_\varepsilon = F_\varepsilon^q, \text{ in } \Omega, \tag{4.98}$$

$$\nabla q_\varepsilon \cdot \nu = 0 \text{ on } \Gamma,$$

where

$$F_\varepsilon^q(x) = -y_\varepsilon^*(0, x)p_\varepsilon(0, x)$$
$$- \int_0^T \left(p_\varepsilon \frac{dy_\varepsilon^*}{dt} - \chi_{\Omega_{obs}} \chi_{(0, T_{obs})}(t) y_\varepsilon^* S_\varepsilon^* \right) dt + \frac{1}{\varepsilon} \chi_{\Omega_0} u_\varepsilon^*, \tag{4.99}$$

and $F_\varepsilon^q \in L^2(\Omega)$. We remark that (4.98) has a unique solution $q_\varepsilon \in H^2(\Omega)$.

4.5.1.2   The Optimality Condition

Proposition 4.11. *Assume (4.5), (4.88) and let v_ε^* be an optimal control in (P_ε). Then,*

$$\begin{cases} v_\varepsilon^* = \phi_{\min} & \text{on } \{x \in \Omega; \ q_\varepsilon(x) > 0\} \\ v_\varepsilon^* \in (\phi_{\min}, u_M) & \text{on } \{x \in \Omega; \ q_\varepsilon = 0\} \\ v_\varepsilon^* = u_M & \text{on } \{x \in \Omega; \ q_\varepsilon(x) < 0\}. \end{cases} \tag{4.100}$$

Proof. We multiply (4.47) by p_ε and integrate over Q, getting again (4.71), i.e.,

$$\int_Q Y_\varepsilon(u_\varepsilon^* + \varepsilon) S_\varepsilon^* \chi_{(0, T_{obs})}(t) \chi_{\Omega_{obs}}(x) dx dt \tag{4.101}$$

$$= - \int_\Omega p_\varepsilon(0, x) y_\varepsilon^*(0, x) \widetilde{u}_\varepsilon dx - \int_Q p_\varepsilon \frac{dy_\varepsilon^*}{dt} \widetilde{u}_\varepsilon dx dt.$$

Next, we multiply (4.97) by q_ε and integrate over Ω, obtaining

$$\int_\Omega (q_\varepsilon - \Delta q_\varepsilon)\widetilde{u}_\varepsilon dx + \int_\Gamma \widetilde{u}_\varepsilon \nabla q_\varepsilon \cdot \nu d\sigma = \int_\Omega \widetilde{v}_\varepsilon q_\varepsilon dx. \tag{4.102}$$

Therefore

$$\int_\Omega \widetilde{u}_\varepsilon(x) F_\varepsilon^q(x)dx = \int_\Omega \widetilde{v}_\varepsilon(x) q_\varepsilon(x)dx. \tag{4.103}$$

On the other hand, from the fact that v_ε^* is optimal in $(\widetilde{P}_\varepsilon)$ we deduce that

$$\int_{Q_{obs}} ((u_\varepsilon^* + \varepsilon)Y_\varepsilon + \widetilde{u}_\varepsilon y_\varepsilon^*)S_\varepsilon^* dxdt + \frac{1}{\varepsilon}\int_{\Omega_0} u_\varepsilon^* \widetilde{u}_\varepsilon dx \geq 0. \tag{4.104}$$

We replace (4.101) on the left-hand side and get

$$\int_\Omega \widetilde{u}_\varepsilon \left\{ -p_\varepsilon(0,x)y_\varepsilon^*(0,x) - \int_0^T \left(p_\varepsilon \frac{dy_\varepsilon^*}{dt} - \chi_{\Omega_{obs}}\chi_{(0,T_{obs})}y_\varepsilon^* S_\varepsilon^* \right) dt \right.$$
$$\left. + \frac{1}{\varepsilon}u_\varepsilon^* \chi_{\Omega_0} \right\} dx \geq 0.$$

Taking into account (4.99), (4.103) and recalling that $\widetilde{v}_\varepsilon = w_\varepsilon - v_\varepsilon^*$, we obtain

$$\int_\Omega (v_\varepsilon^* - w_\varepsilon)(-q_\varepsilon) dx \geq 0, \text{ for any } w_\varepsilon \in \widetilde{W} \tag{4.105}$$

which implies that

$$-q_\varepsilon \in \partial I_{K_1}(v_\varepsilon^*) = N_{K_1}(v_\varepsilon^*) \tag{4.106}$$

where by $\partial I_{K_1}(v_\varepsilon^*)$ we denote the subdifferential of the indicator function of the closed set $K_1 = [\phi_{\min}(x), u_M]$ at v_ε^*.

We recall that the indicator function of a closed convex subset $\mathcal{K}$ of a Banach space X is a function $I_\mathcal{K} : X \to (-\infty, \infty]$

$$I_\mathcal{K}(x) = \begin{cases} 0 & \text{if } x \in \mathcal{K}, \\ +\infty & \text{otherwise} \end{cases}$$

and

$$\partial I_\mathcal{K}(x) = \begin{cases} \{0\} & \text{if } x \in \text{int}\mathcal{K}, \\ N_\mathcal{K}(x) & \text{if } x \in \partial\mathcal{K}. \end{cases}$$

Here, $N_\mathcal{K}(x) \subset X'$ is the normal cone to $\mathcal{K}$ at x and is defined by

$$N_\mathcal{K}(x) = \{x^* \in X'; \langle x^*, x - y \rangle_{X',X} \geq 0, \ \forall y \in \mathcal{K}\}.$$

In conclusion by (4.106) we get the optimality condition (4.100), as claimed.
$\square$

4.5.2   Numerical Results

We present numerical results for identifying the function u_ε^*, a solution to $(\widetilde{P}_\varepsilon)$, by following the alternative approach and several steps based on Rosen's algorithm (see [5]). The space domain is $\Omega = \{(x_1, x_2); x_1 \in (0, 5), x_2 \in (0, 5)\}$, $\Omega_0 = (2, 3) \times (2, 3)$ and $\Omega_{obs} = (3, 5) \times (3, 5)$.

We apply the method for the following data

$$\beta^*(r) = r^2, \quad K_0 = 0, \quad f(t, x_1, x_2) = \begin{cases} 0.1x_1 \text{ in } \Omega_u \\ 0 \quad \text{ in } \Omega_0 \end{cases},$$

$$\theta_0 = \begin{cases} 0.98 \text{ in } \Omega_u \\ 0 \quad \text{ in } \Omega_0 \end{cases}, \quad \theta_{obs} = 0.5.$$

The control is v_ε^* and u_ε^* will be determined by (4.9). For simplicity we shall not retain the subscript ε for the control and state.

Step 0.   Fix ε_{crt} small and let us take $v_0^*(x) \in [\phi_{\min}, u_M]$. Set $k = 0$.

Step 1.   We determine u_k^* by (4.9), y_k^* by (4.15)–(4.17), p_k by (4.60)–(4.62), q_k by (4.98) and $\widetilde{J}_\varepsilon(v_k^*)$ from (4.95).

Step 2.   We compute w_k by (4.100)

$$\begin{cases} w_k = \phi_{\min} & \text{on } \{x \in \Omega; \ q_k(x) > 0\} \\ w_k \in (\phi_{\min}, u_M) & \text{on } \{x \in \Omega; \ q_k = 0\} \\ w_k = u_M & \text{on } \{x \in \Omega; \ q_k(x) < 0\}. \end{cases}$$

If $q_k = 0$ we can try with w_k any value in $(\phi_{\min}, u_M)$. Then we compute $\widetilde{v}_k = w_k - v_k^*$ and

$$v_k(x) = v_k^*(x) + \lambda \widetilde{v}_k(x) = (1 - \lambda_k)v_k^*(x) + \lambda w_k(x), \ \lambda \in [0, 1]$$

such that
$$\widetilde{J}_\varepsilon(v_k) = \min_{\lambda \in [0,1]} \{\widetilde{J}_\varepsilon(v_k^* + \lambda \widetilde{v}_k(x))\}. \tag{4.107}$$

Step 3.   We set
$$v_{k+1}^*(x) = v_k(x)$$

with v_k found by (4.107) and compute $\widetilde{J}_\varepsilon(v_{k+1}^*)$. Denote

$$err = \left| \widetilde{J}_\varepsilon(v_{k+1}) - \widetilde{J}_\varepsilon(v_k) \right|.$$

If $\widetilde{J}_\varepsilon$ decreases and $err \leq \varepsilon_{crt}$ with the prescribed ε_{crt} then the algorithm stops and set $v_\varepsilon^* = v_{k+1}^*$ and u_{k+1}^* is obtained from (4.9).
If not, continue from Step 1.

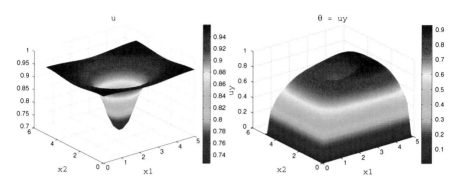

Fig. 4.1 The approximating control u_ε^* and state function θ_ε^*

In Fig. 4.1 we see the graphics of u_ε^* and the corresponding $\theta_\varepsilon^* = u_\varepsilon^* y_\varepsilon^*$ computed according the above algorithm, by Comsol + Matlab, with

$$v_0^* = u_M, \quad \varepsilon_{crt} = 0.1,$$

after three steps, in which the obtained values are indicated below:

Step 0: $\widetilde{J}_\varepsilon(v_0) = 12.3691$
Step 1: $\widetilde{J}_\varepsilon(v_1) = 9.2221$
Step 2: $\widetilde{J}_\varepsilon(v_2) = 9.194.$

References

1. S. Agmon, A. Douglis, L. Nirenberg, Estimates near boundary for solutions of elliptic partial differential equations satisfying general boundary conditions. Comm. Pure Appl. Math. **12**, 623–727 (1959)
2. M. Al Horani, A. Favini, An identification problem for first-order degenerate differential equations. J. Optim. Theor. Appl. **130**(1), 41–60 (2006)
3. M. Al Horani, A. Favini, A. Lorenzi, Second-order degenerate identification differential problems. J. Optim. Theor. Appl. **141**(1), 13–36 (2009)
4. H.W. Alt, S. Luckhaus, Quasi-linear elliptic-parabolic differential equations. Math. Z. **183**, 311–341 (1983)
5. V. Arnăutu, P. Neittaanmäki, *Optimal Control from Theory to Computer Programs* (Kluwer Academic, Dordrecht, 2003)
6. D.G. Aronson, The porous medium equation, in *Some Problems in Nonlinear Diffusion*, ed. by A. Fasano, M. Primicerio. Lecture Notes in Mathematics, vol 1224 (Springer, Berlin, 1986)
7. J.P. Aubin, Un théorème de compacité. C. R. Acad. Sci. Paris Sér. I **256**, 5042–5044 (1963)
8. T. Barbu, Eigenimage-based face recognition approach using gradient covariance. Numer. Func. Anal. Optim. **28**(5–6), 591–601 (2007)
9. T. Barbu, V. Barbu, V. Biga, D. Coca, A PDE variational approach to image denoising and restoration. Nonlin. Anal. R. World Appl. **10**(3), 1351–1361 (2009)
10. V. Barbu, *Nonlinear Semigroups and Differential Equations in Banach Spaces* (Editura Academiei-Noordhoff International Publishing, Bucureşti-Leyden, 1976)
11. V. Barbu, *Analysis and Control of Nonlinear Infinite Dimensional Systems* (Academic, New York, 1993)
12. V. Barbu, *Mathematical Methods in Optimization of Differential Systems* (Kluwer Academic, Dordrecht, 1994)
13. V. Barbu, *Partial Differential Equations and Boundary Value Problems* (Kluwer Academic, Dordrecht, 1998)
14. V. Barbu, *Nonlinear Differential Equations of Monotone Type in Banach Spaces* (Springer, New York, 2010)
15. V. Barbu, Self-organized criticality and convergence to equilibrium of solutions to nonlinear diffusion equations. Annu. Rev. Contr. **34**(1), 52–61 (2010)
16. V. Barbu, A. Favini, Periodic problems for degenerate differential equations. Rend. Istit. Mat. Univ. Trieste Suppl. **XXVIII**, 29–57 (1997)
17. V. Barbu, A. Favini, Periodic solutions to degenerate second order differential equations in Hilbert spaces. Commun. Appl. Anal. **2**, 19–29 (1998)

A. Favini and G. Marinoschi, *Degenerate Nonlinear Diffusion Equations*, 135
Lecture Notes in Mathematics 2049, DOI 10.1007/978-3-642-28285-0,
© Springer-Verlag Berlin Heidelberg 2012

18. V. Barbu, A. Favini, Control of degenerate differential systems. Recent advances in control of PDEs. Contr. Cybern. **28**(3), 397–420 (1999)

19. V. Barbu, A. Favini, Analytic semigroups on $H^1(\Omega)$ generated by degenerate elliptic operators. Semigroup Forum **62**(2), 317–328 (2001)

20. V. Barbu, G. Marinoschi, Existence for a time dependent rainfall infiltration model with a blowing up diffusivity. Nonlin. Anal. R. World Appl. **5**(2), 231–245 (2004)

21. V. Barbu, T. Precupanu, *Convexity and Optimization in Banach Spaces* (D. Reidel, Dordrecht, 1986)

22. V. Barbu, T. Precupanu, *Convexity and Optimization in Banach Spaces* (Springer, Dordrecht, Heidelberg, London, New York, 2012)

23. P. Bénilan, S.N. Krushkov, Quasilinear first-order equations with continuous nonlinearities. Russian Acad. Sci. Dokl. Math. **50**(3), 391–396 (1995)

24. I. Borsi, A. Farina, A. Fasano, On the infiltration of rain water through the soil with runoff of the excess water. Nonlin. Anal. R. World Appl. **5**, 763–800 (2004)

25. I. Borsi, A. Farina, R. Gianni, M. Primicerio, Continuous dependence on the constitutive functions for a class of problems describing fluid flow in porous media. Atti Accad. Naz. Lincei Cl. Sci. Fis. Mat. Natur. Rend. Lincei 9 Mat. Appl. **20**(1), 1–24 (2009)

26. H. Brezis, Equations et inéquations nonlinéaires dans les espaces vectorielles en dualité. Ann. Inst. Fourier **18**, 115–175 (1968)

27. H. Brezis, Monotonicity methods in Hilbert spaces and some applications to nonlinear partial differential equations, in *Contributions to Nonlinear Functional Analysis*, ed. by E. Zarantonello (Academic, New York, 1971)

28. H. Brezis, Problemes unilatéraux. J. Math. Pure. Appl. **51**, 1–168 (1972)

29. H. Brezis, *Opérateurs Maximaux Monotones et Semi-groupes de Contractions dans les Espaces de Hilbert* (North-Holland, Amsterdam, 1973)

30. H. Brezis, *Functional Analysis, Sobolev Spaces and Partial Differential Equations* (Springer, Berlin, 2011)

31. H. Brezis, I. Ekeland, Un principe variationnel associé à certaines equations paraboliques. Le cas independant du temps. C. R. Acad. Sci. Paris Sér. A **282**, 971–974 (1976)

32. H. Brezis, I. Ekeland, Un principe variationnel associé à certaines equations paraboliques. Le cas dependant du temps. C. R. Acad. Sci. Paris Sér. A **282**, 1197–1198 (1976)

33. P. Broadbridge, I. White, Constant rate rainfall infiltration: A versatile nonlinear model, 1. Analytic solution. Water Resour. Res. **24**(1), 145–154 (1988)

34. F. Browder, Problèmes Nonlinéaires, Les Presses de l'Université de Montréal, 1966

35. F. Browder, Nonlinear operators and nonlinear equations of evolution in Banach spaces, in *Nonlinear Functional Analysis* ed. by F. Browder. Symposia in Pure Math, vol 18, Part 2 (American Mathematical Society, Providence, 1970)

36. S. Busenberg, M. Iannelli, Degenerate nonlinear diffusion problem in age-structured population dynamics. Nonlinear Anal. Theor. Meth. Appl. **7**(12), 1411–1429 (1983)

37. J. Carillo, Entropy solutions for nonlinear degenerate problems. Arch. Ration. Mech. Anal. **147**, 269–361 (1999)

38. J. Carillo, P. Wittbold, Uniqueness of renormalized solutions of degenerate elliptic-parabolic problems. J. Differ. Equat. **156**, 93–121 (1999)

39. C. Ciutureanu, G. Marinoschi, Convergence of the finite difference scheme for a fast diffusion equation in porous media. Numer. Func. Anal. Optim. **29**, 1034–1063 (2008)

40. COMSOL Multiphysics v3.5a. Floating Network License 1025226, Comsol Sweden (2007)

41. M.G. Crandall, A. Pazy, Semigroups of nonlinear contractions and dissipative sets. J. Funct. Anal. **3**, 376–418 (1969)

42. M.G. Crandall, A. Pazy, On accretive sets in Banach spaces. J. Funct. Anal. **5**, 204–217 (1970)

43. M.G. Crandall, T.M. Liggett, Generation of semigroups of nonlinear transformations in general Banach spaces. Amer. J. Math. **93**, 265–298 (1971)

44. M.G. Crandall, A. Pazy, Nonlinear evolution equations in Banach spaces. Isr. J. Math. **11**, 57–94 (1971)

45. M.G. Crandall, L.C. Evans, On the relation of the operator $\partial/\partial s + \partial/\partial t$ to evolution governed by accretive operators. Isr. J. Math. **21**, 261–278 (1975)

46. M.G. Crandall, A. Pazy, An approximation of integrable functions by step functions with an application. Proc. Amer. Math. Soc. **76**(1), 74–80 (1979)

47. C.M Dafermos, M. Slemrod, Asymptotic behavior of nonlinear contraction semigroups. J. Funct. Anal. **13**, 97–106 (1973)

48. E. DiBenedetto, R.E. Showalter, Implicit degenerate evolution equations and applications. SIAM J. Math. Anal. **12**, 731–751 (1981)

49. A. Fasano, M. Primicerio, Free boundary problems for nonlinear parabolic equations with nonlinear free bounadry conditions. J. Math. Anal. Appl. **72**, 247–273 (1979)

50. A. Fasano, M. Primicerio, Liquid flow in partially saturated porous media. J. Inst. Math. Appl. **23**, 503–517 (1979)

51. A. Favini, The regulator problem for a singular control system, in *Evolution Equations*. Lecture Notes in Pure and Applied Mathematics, vol 234 (Dekker, New York, 2003), pp. 191–201

52. A. Favini, M. Fuhrman, Approximation results for semigroups generated by multivalued linear operators and applications. Differ. Integr. Equat. **11**(5), 781–805 (1998)

53. A. Favini, A. Lorenzi, Singular integro-differential equations of parabolic type and inverse problems. Math. Model. Meth. Appl. Sci. **13**(12), 1745–1766 (2003)

54. A. Favini, A. Lorenzi, Identification problems for singular integro-differential equations of parabolic type II. Nonlinear Anal. **56**(6), 879–904 (2004)

55. A. Favini, A. Lorenzi, Identification problems for singular integro-differential equations of parabolic type I. Dyn. Contin. Discrete Impuls. Syst. Ser. A Math. Anal. **12**(3–4), 303–328 (2005)

56. A. Favini, A. Lorenzi, H. Tanabe, A. Yagi, An L^p-approach to singular linear parabolic equations in bounded domains. Osaka J. Math. **42**(2), 385–406 (2005)

57. A. Favini, A. Lorenzi, H. Tanabe, A. Yagi, An L^p-approach to singular linear parabolic equations with lower order terms. Discrete Contin. Dyn. Syst. **22**(4), 989–1008 (2008)

58. A. Favini, G. Marinoschi, Existence for a degenerate diffusion problem with a nonlinear operator. J. Evol. Equ. **7**, 743–764 (2007)

59. A. Favini, G. Marinoschi, Periodic behavior for a degenerate fast diffusion equation. J. Math. Anal. Appl. **351**(2), 509–521 (2009)

60. A. Favini, G. Marinoschi, Identifcation of the time derivative coefficient in a fast diffusion degenerate equation. J. Optim. Theor. Appl. **145**, 249–269 (2010)

61. A. Favini, G. Marinoschi, Identification for degenerate problems of hyperbolic type, in PDEs and Inverse Problems, ed. by A. Favini, A. Lorenzi. Applicable Analysis (online, DOI:10.1080/00036811.2011.630665)

62. A. Favini, A. Yagi, *Degenerate Differential Equations in Banach Spaces* (Marcel Dekker, New York, 1999)

63. A. Favini, A. Yagi, Quasilinear degenerate evolution equations in Banach spaces. J. Evol. Equ. **4**, 421–449 (2004)

64. A. Gandolfi, M. Iannelli, G. Marinoschi, An age-structured model of epidermis growth. J. Math. Biol. **62**(1), 111–141 (2011)

65. J.A. Goldstein, C.Y. Lin, Degenerate nonlinear parabolic boundary problems, in *Nonlinear Analysis and Applications* (Arlington, TX, 1986). Lecture Notes in Pure and Applied Mathematics, vol 109 (Dekker, New York, 1987), pp. 189–196

66. J.A. Goldstein, C.Y. Lin, An L^p-semigroup approach to degenerate parabolic boundary value problems. Ann. Mat. Pura Appl. **159**(4), 211–227 (1991)

67. A. Granas, J. Dugundji, *Fixed Point Theory* (Springer, New York, 2003)

68. A. Haraux, in *Nonlinear Evolution Equations-Global Behaviour of Solutions*. Lecture Notes in Mathematics, vol 841 (Springer, Berlin, 1981)

69. T. Hillen, K.J. Painter, A user's guide to PDE models for chemotaxis. J. Math. Biol. **58**, 183–217 (2009)

70. J. Kačur, S. Luckhaus, Approximation of degenerate parabolic systems by nondegenerate elliptic and parabolic systems. Appl. Numer. Math. **26**, 307–326 (1998)

71. T. Kato, Nonlinear semi-groups and evolution equations. J. Math. Soc. Jpn. **19**, 508–520 (1967)

72. T. Kato, Accretive operators and nonlinear evolution equations in Banach spaces, in *Nonlinear Functional Analysis* ed. by F. Browder (American Mathemathical Society, Providence, 1970), pp. 138–161

73. Y. Kobayashi, Difference approximation of Cauchy problem for quasi-dissipative operators and generation of nonlinear semigroups. J. Math. Soc. Jpn. **27**, 641–663 (1975)

74. Y. Komura, Nonlinear semigroups in Hilbert spaces. J. Math. Soc. Jpn. **19**, 508–520 (1967)

75. S.N. Krushkov, Generalized solutions of the Cauchy problem in the large for first-order nonlinear equations. Sov. Math. Dokl. **10**, 785–788 (1969)

76. S.N. Krushkov, First order quasilinear equations in several independent variables. Mat. Sb. **81**, 228–255 (1970)

77. J.L. Lions, *Quelques Méthodes de Résolution des Problèmes aux Limites non Linéaires* (Dunod, Paris, 1969)

78. J. L. Lions, E. Magenes, *Non-homogeneous Boundary Value Problems and Applications, I* (Springer, Berlin, 1972)

79. G. Minty, Monotone (nonlinear) operators in Hilbert spaces. Duke Math. J. **29**, 341–346 (1962)

80. G. Minty, On the generalization of a direct method of the calculus of variations. Bull. Amer. Math. Soc. **73**, 315–321 (1967)

81. G. Marinoschi, Nonlinear infiltration with a singular diffusion coefficient. Differ. Integr. Equat. **16**(9), 1093–1110 (2003)

82. G. Marinoschi, A free boundary problem describing the saturated unsaturated flow in a porous medium. Abstr. Appl. Anal. **9**, 729–755 (2004)

83. G. Marinoschi, A free boundary problem describing the saturated unsaturated flow in a porous medium, II. Existence of the free boundary in the 3-D case. Abstr. Appl. Anal. **8**, 813–854 (2005)

84. G. Marinoschi, Functional approach to nonlinear models of water flow in soils. *Mathematical Modelling: Theory and Applications*, vol. 21 (Springer, Dordrecht, 2006)

85. G. Marinoschi, A hysteresis model for an infiltration-drainage process. Nonlinear Anal. R. World Appl. **9**, 518–535 (2008)

86. G. Marinoschi, Nonlinear diffusion equations with discontinuous coefficients in porous media, in *Progress in Nonlinear Analysis Research*, ed. by Erik T. Hoffmann (Nova Science, New York, 2008), pp. 209–242

87. G. Marinoschi, Periodic solutions to fast diffusion equations with non Lipschitz convective terms. Nonlinear Anal. R. World Appl. **10**, 1048–1067 (2009)

88. G. Marinoschi, Well posedness of a time-difference scheme for a degenerate fast diffusion problem. Discrete Continuous Dyn. Syst. B **13**, 435–454 (2010)

89. G. Marinoschi, Well-posedness of singular diffusion equations in porous media with homogeneous Neumann boundary conditions. Nonlinear Anal. Theor. Meth. Appl. **72**, 3491–3514 (2010)

90. G. Marinoschi, Existence to time-dependent nonlinear diffusion equations via convex optimization. JOTA **154**, 3 (2012) (online, DOI: 10.1007/s10957-012-0017-6)

91. J. Nečas, *Les méthodes directes en théorie des équations elliptiques* (Masson, Paris-Academia, Praha, 1967)

92. F. Otto, L^1-contraction and uniqueness for unstationary saturated-unsaturated porous media flow. Adv. Math. Sci. Appl. **7**, 537–553 (1997)

93. R.T. Rockafellar, *Convex Analysis* (Princeton University Press, Princeton, 1969)

94. J. Rulla, R.E. Showalter, Diffusion in partially fissured media and implicit evolution systems. Adv. Math. Sci. Appl. **5**(1), 163–191 (1995)

95. R.E. Showalter, Monotone operators in Banach spaces and nonlinear partial differential equations, in *Mathematical Surveys and Monographs*, vol 49 (AMS, Providence, 1997)

96. J.L. Vázquez, Darcy's law and the theory of shrinking solutions of fast diffusion equations. SIAM J. Math. Anal. **35**(4), 1005–1028 (2003)

97. J.L. Vázquez, Smoothing and decay estimates for nonlinear diffusion equations, in *Equations of Porous Medium Type*. Series Oxford Lecture Series in Mathematics and its Applications, vol 33 (Oxford University Press, Oxford, 2006)

98. J. L. Vázquez, The porous medium equation. Mathematical Theory (Oxford Mathematical Monographs. The Clarendon Press, Oxford University Press, Oxford, 2007)

99. K. Yosida, *Functional Analysis* (Springer, Berlin, 1980)

List of Symbols

N	Space dimension
$\mathbb{N}^*$	The set of natural numbers without 0
$\mathbb{R}$	The set of real numbers
$\mathbb{R}^N$	N-dimension real set
supp f	Support of function f
$\mathcal{D}(\Omega)$	Space of distributions on Ω
$\mathcal{D}'(\Omega)$	Dual of $\mathcal{D}(\Omega)$
$L^p(\Omega)$	$\{y : \Omega \to \mathbb{R};\ y$ measurable, $\left(\int_\Omega |y(x)|^p\, dx\right)^{1/p} < \infty\},\ p \in [1, \infty)\}$
$L^\infty(\Omega)$	$\{y : \Omega \to \mathbb{R};\ y$ measurable, $\exists C > 0$ such that $|y(x)| \leq ess \sup_{x\in\Omega} |y(x)|)$
$L^p(0, T; X)$	$\{y : (0, T) \to X;\ y$ measurable, $\|y(t)\|_X^p$ is Lebesgue integrable over $(0, T)\},\ p \in [1, \infty)\}$
$L^\infty(0, T; X)$	$\{y : (0, T) \to X;\ y$ measurable, $ess \sup_{t\in(0,T)} \|y(t)\| < \infty\}$
$W^{1,2}(\Omega),\ W^{2,2}(\Omega),$ $H^1(\Omega),\ H_0^1(\Omega),$ $H^2(\Omega)$	Sobolev spaces
$W^{k,p}([0,T]; X)$	$\left\{y \in \mathcal{D}'(0, T; X);\ \frac{d^j y}{dt^j} \in L^p(0, T; X),\ j = 0, ..., k\right\},$ $p = 1, 2, ...$
$\|\ \|, (\cdot, \cdot)$	Norm and scalar product on $L^2(\Omega)$
$\to$	Strong convergence
$\rightharpoonup$	Weak convergence

A. Favini and G. Marinoschi, *Degenerate Nonlinear Diffusion Equations*,
Lecture Notes in Mathematics 2049, DOI 10.1007/978-3-642-28285-0,
© Springer-Verlag Berlin Heidelberg 2012

$\overset{w*}{\rightarrow}$ Weak-star convergence

∇u $\left(\frac{\partial u}{\partial x_1}, \frac{\partial u}{\partial x_2}, ..., \frac{\partial u}{\partial x_N} \right)$ gradient of the scalar
function $u(x_1, ..., x_N)$

Δy $\sum\limits_{i=1}^{N} \frac{\partial^2 y}{\partial x_i^2}$ Laplacian of $y(x_1, ..., x_N)$

$\nabla \cdot a$ $\sum\limits_{i=1}^{N} \frac{\partial a_i}{\partial x_i}$ the divergence of the vector
$a = (a_1, ... a_N)$

$C(\Omega)$ The space of all continuous real valued functions
on Ω

$C^k(\Omega)$ The space of all continuously differentiable
functions on Ω of order $m \leq k$

$C_0^\infty(\Omega)$ The space of indefinitely differentiable functions
with compact support in Ω

$BV([0,T]; X)$ The space of bounded variation functions from
$[0,T]$ to X

Index

degenerate
    parabolic-elliptic, xviii
diffusion
    coefficient, xiii
    degenerate, xix
    fast, xiv
    nondegenerate, xix
    slow, xv
    superdiffusion, xv
    very fast, xv

Fatou, 11
function
    convex, 10
    lower semicontinuous, 10
    proper, 10
    weakly lower semicontinuous, 10

indicator function, 131

Lebesgue point, 33

multiplicator, 7

normal cone, 131

operator
    coercive, 16
    demiclosed, 22
    monotone, 14
    quasi m-accretive, 14

Poincaré inequality, 6

solution
    generalized, 7, 45, 60
    h-approximate, 62
    mild, 9, 61, 62
    strong, 9
    weak, 7
subdifferential, 10

theorem
    Ascoli-Arzelà, 21
    Aubin, Lions, 20
    Minty, 14
    Schauder-Tychonoff, 50

A. Favini and G. Marinoschi, *Degenerate Nonlinear Diffusion Equations*,
Lecture Notes in Mathematics 2049, DOI 10.1007/978-3-642-28285-0,
© Springer-Verlag Berlin Heidelberg 2012

LECTURE NOTES IN MATHEMATICS   ⟲ Springer

Edited by J.-M. Morel, B. Teissier; P.K. Maini

Editorial Policy (for the publication of monographs)

1. Lecture Notes aim to report new developments in all areas of mathematics and their applications - quickly, informally and at a high level. Mathematical texts analysing new developments in modelling and numerical simulation are welcome.

   Monograph manuscripts should be reasonably self-contained and rounded off. Thus they may, and often will, present not only results of the author but also related work by other people. They may be based on specialised lecture courses. Furthermore, the manuscripts should provide sufficient motivation, examples and applications. This clearly distinguishes Lecture Notes from journal articles or technical reports which normally are very concise. Articles intended for a journal but too long to be accepted by most journals, usually do not have this "lecture notes" character. For similar reasons it is unusual for doctoral theses to be accepted for the Lecture Notes series, though habilitation theses may be appropriate.

2. Manuscripts should be submitted either online at www.editorialmanager.com/lnm to Springer's mathematics editorial in Heidelberg, or to one of the series editors. In general, manuscripts will be sent out to 2 external referees for evaluation. If a decision cannot yet be reached on the basis of the first 2 reports, further referees may be contacted: The author will be informed of this. A final decision to publish can be made only on the basis of the complete manuscript, however a refereeing process leading to a preliminary decision can be based on a pre-final or incomplete manuscript. The strict minimum amount of material that will be considered should include a detailed outline describing the planned contents of each chapter, a bibliography and several sample chapters.

   Authors should be aware that incomplete or insufficiently close to final manuscripts almost always result in longer refereeing times and nevertheless unclear referees' recommendations, making further refereeing of a final draft necessary.

   Authors should also be aware that parallel submission of their manuscript to another publisher while under consideration for LNM will in general lead to immediate rejection.

3. Manuscripts should in general be submitted in English. Final manuscripts should contain at least 100 pages of mathematical text and should always include

   – a table of contents;
   – an informative introduction, with adequate motivation and perhaps some historical remarks: it should be accessible to a reader not intimately familiar with the topic treated;
   – a subject index: as a rule this is genuinely helpful for the reader.

   For evaluation purposes, manuscripts may be submitted in print or electronic form (print form is still preferred by most referees), in the latter case preferably as pdf- or zipped psfiles. Lecture Notes volumes are, as a rule, printed digitally from the authors' files. To ensure best results, authors are asked to use the LaTeX2e style files available from Springer's web-server at:

   ftp://ftp.springer.de/pub/tex/latex/svmonot1/ (for monographs) and
   ftp://ftp.springer.de/pub/tex/latex/svmultt1/ (for summer schools/tutorials).

Additional technical instructions, if necessary, are available on request from lnm@springer.com.

4. Careful preparation of the manuscripts will help keep production time short besides ensuring satisfactory appearance of the finished book in print and online. After acceptance of the manuscript authors will be asked to prepare the final LaTeX source files and also the corresponding dvi-, pdf- or zipped ps-file. The LaTeX source files are essential for producing the full-text online version of the book (see http://www.springerlink.com/openurl.asp?genre=journal&issn=0075-8434 for the existing online volumes of LNM). The actual production of a Lecture Notes volume takes approximately 12 weeks.

5. Authors receive a total of 50 free copies of their volume, but no royalties. They are entitled to a discount of 33.3 % on the price of Springer books purchased for their personal use, if ordering directly from Springer.

6. Commitment to publish is made by letter of intent rather than by signing a formal contract. Springer-Verlag secures the copyright for each volume. Authors are free to reuse material contained in their LNM volumes in later publications: a brief written (or e-mail) request for formal permission is sufficient.

Addresses:

Professor J.-M. Morel, CMLA,
École Normale Supérieure de Cachan,
61 Avenue du Président Wilson, 94235 Cachan Cedex, France
E-mail: morel@cmla.ens-cachan.fr

Professor B. Teissier, Institut Mathématique de Jussieu,
UMR 7586 du CNRS, Équipe "Géométrie et Dynamique",
175 rue du Chevaleret
75013 Paris, France
E-mail: teissier@math.jussieu.fr

For the "Mathematical Biosciences Subseries" of LNM:

Professor P. K. Maini, Center for Mathematical Biology,
Mathematical Institute, 24-29 St Giles,
Oxford OX1 3LP, UK
E-mail : maini@maths.ox.ac.uk

Springer, Mathematics Editorial, Tiergartenstr. 17,
69121 Heidelberg, Germany,
Tel.: +49 (6221) 4876-8259

Fax: +49 (6221) 4876-8259
E-mail: lnm@springer.com